The Cosmic Evolution of Galaxy Structure

AMERICAN ASTRONOMICAL SOCIETY IOP | ebooks™

About the program:

AAS-IOP Astronomy ebooks is the official book program of the American Astronomical Society (AAS), and aims to share in depth the most fascinating areas of astronomy, astrophysics, solar physics and planetary science. The program includes publications in the following topics:

Books in the program range in level from short introductory texts on fast-moving areas, graduate and upper-level undergraduate textbooks, research monographs and practical handbooks.

For a complete list of published and forthcoming titles, please visit iopscience.org/books/aas.

About the American Astronomical Society

The American Astronomical Society (aas.org), established 1899, is the major organization of professional astronomers in North America. The membership (~7,000) also includes physicists, mathematicians, geologists, engineers and others whose research interests lie within the broad spectrum of subjects now comprising the contemporary astronomical sciences. The mission of the Society is to enhance and share humanity's scientific understanding of the universe.

Editorial Advisory Board

The Cosmic Evolution of Galaxy Structure

Christopher J Conselice
Jodrell Bank Centre for Astrophysics
University of Manchester, Manchester, UK

IOP Publishing, Bristol, UK

ISBN 978-0-7503-2670-4 (ebook)
ISBN 978-0-7503-2668-1 (print)
ISBN 978-0-7503-2671-1 (myPrint)
ISBN 978-0-7503-2669-8 (mobi)

DOI 10.1088/2514-3433/abb602

Version: 20201201

AAS–IOP Astronomy
ISSN 2514-3433 (online)
ISSN 2515-141X (print)

British Library Cataloguing-in-Publication Data: A catalogue record for this book is available from the British Library.

Published by IOP Publishing, wholly owned by The Institute of Physics, London

IOP Publishing, Temple Circus, Temple Way, Bristol, BS1 6HG, UK

US Office: IOP Publishing, Inc., 190 North Independence Mall West, Suite 601, Philadelphia, PA 19106, USA

Dedicated to my former and current PhD students and postdocs who have contributed greatly to my understanding of the material in this book.

Contents

Preface

It is often said that astronomy is the oldest of the sciences. This is probably true as it is quite unavoidable to wonder about the stars that must have been very visible at night to early humans. This is especially the case as the night sky, until the last few hundred years, was extremely dark and this magnificent display could not have been ignored. The importance of the stars, the planets, the Moon, and the Milky Way to early humans cannot be known, but must have had great importance and a major part of how early humans saw their origin and developed their cosmology and religion. Today, the night sky is no different, but we can now probe these objects and ones which are unbelievably fainter to scientifically answer some of the questions that humans have always had about themselves and the universe—including: where did we come from and where are we going?

We know that in one way or another astronomers have been cataloging objects in the sky for thousands of years, and the discovery of planets such as Mars and Jupiter were not discovered in historic times, and we associate no known person to their discoveries. It is for certain that people noticed the wanderings of these planets among the bright stars long before written records were made. This further shows how attentive early humans were to the night sky.

Early astronomers about 2000 years ago, including Hipparchus and Ptolemy, started to catalog the objects that we see in the sky, especially the stars. By the time of these astronomers there was a distinct interest in cataloging stars as well as mapping the motions of the planets. Mathematical (but not physical) models of the motions of these planets were created, which included the famous epicycles of planetary motion to explain why there was retrograde motion seen on the night sky for these moving objects, as famously developed by the 2nd-century Greek/Roman astronomer Claudius Ptolemy. In his book "The Almagest" he sets out to describe his Geocentric view of the universe which dominated astronomical thought for over 1000 years.

Within the Almagest there was a star catalog of about 1020 objects, those which can be observed from Alexandria, Egypt where Ptolemy spend a large fraction of his life at its famous library. This was one of the first star catalogs, after earlier work by astronomers such as Timocharis of Alexandria, Aristyllus, and of course Hipparchus who took much of the data Ptolemy used in his work, hundreds of years earlier.

During this cataloging and observations it became known to some careful observers that not all objects in the sky were planets or stars. In particular, Islamic astronomers noticed that there were several patches of the night sky which looked different from stars, and were made up of "fuzzy" or "diffuse" light. These objects were diffuse and extended, and in the language of today we would call them low surface brightness objects. It was the Persian astronomer Abd al-Rahman al-Sufi, who lived in what is now Iran, in his book "Book of Fixed Stars" who first described the Andromeda galaxy as a "little cloud."

If you are at a dark site today you can see what these ancient astronomers also saw. Most famously if you look below the belt of the Orion constellation you will see a fuzzy spot of light which is now known as the Orion nebula, also noted by al-Sufi. Ancient astronomers did not think much of these objects, for the most part, no doubt partially because they were considered to be very rare and therefore not worthy of detailed study. Another good reason is that even if there was an interest in studying them, before the invention of the telescopes in the early 17th century it would have been impossible to learn much more about them.

This is where things stood for galaxies and galaxy structures for centuries. It is worth noting however that it is the resolved nature of these objects that lead them to be discovered before telescopes—already an indication that what we now call galaxies in fact do have some structure that can be observed and studied. Before the epoch of telescopic astronomy work on cataloging, the study of the positions of stars reached new heights with the extremely technical and detailed observations and derivations of European astronomers, including the Danish astronomer Tycho Brahe, and later Johannes Kepler. It was Kepler who fitted the data that Brahe obtained to develop, in basic terms, a theory of planetary motion that we still use today—one with elliptical, as opposed to circular, orbits.

Soon after this the telescope was invented and used by Galileo Galilei to observe objects in the universe for the first time at high magnification, forever changing astronomy. However, for whatever reason Galileo did not spend much time on the nebula, as it would again take some time before telescopes improved enough to study them in detail. In fact, for the most part, the development of astronomy and cosmology up into the late 18th century did not involve galaxies in a significant way whatsoever.

As detailed in Chapter 1 of this book, the study of galaxies really took off in the 18th century by astronomers searching for comets. It is perhaps ironic to think that the development of the field of galaxy studies started with a byproduct of the search for extreme local objects—comets from our own solar system. The development of photography in the 19th century changed the field of galaxy studies even further, and made it one of the leading areas of all astronomy in the 20th century. This book gives the background to this transformation and explains why the use of galaxy structure has been critical for the understanding of galaxies, and continues to do so.

As recounted throughout the start of this book these searches led to the first cataloging of nebula, many of which we would today call galaxies. These early extragalactic observers had no idea that these were galaxies, something that was not developed or convincingly known until the early 1920s. Some basic spectroscopic work predated detailed imaging of galaxies. Without spectra and the subsequent development of electronic imaging devices our knowledge of galaxies would be blunted a significant amount. This early work was however very difficult and only showed that the spectra of galaxies was in some ways quite similar to that of stars.

However, the early studies of galaxies using photographic plates which imaged their structures really began the field of the astrophysical study of galaxies. This marks the first time data on these systems was obtained beyond simply measuring the flux coming from these systems. One of the first things done was to classify different types of galaxies based on their morphology or apparent structure as judged by the eye at visual wavelengths where early 20th-century telescopes operate at.

This type of work began the study of galaxy structure and morphology which in some ways is the first property of galaxies ever studied. However, just because these structures and morphologies have been studied for longer than any other aspect does not mean that there is not new and interesting information that can be derived today from their structures. Quite the contrary, galaxy structure and morphology offers the most fundamental information about galaxies, and continues to evolve and expand over time.

Today we know there are about 2 trillion galaxies in the universe which we in principle can observe. Each of these galaxies has a formation history, including the formation of stars, dark matter, gas, etc. All of these formation mechanisms have an effect on the structure of a galaxy—its size, its overall morphology, its structure, whether it has asymmetric light, if it has clumpy light features due to star formation episodes, among others. The history and structure of a galaxy can also have an influence on the internal motions of a galaxy and how fast it is rotating or how fast the internal motions are.

Galaxy structure has also proven to be a major way in which the formation of galaxies can be studied in detail. In the universe today there are in general two different ways in which objects form. This is in general either through accretion of material—usually gas, or through the mergers of two existing objects to make a larger object. Astronomers have debated which of the two are the correct ways in which objects are formed. This is the case not only for galaxies, but for the formation of black holes, planets and stars themselves. Through structure we can help answer this question for galaxies and we demonstrate throughout this book that galaxy mergers are a key aspect for forming galaxies.

Galaxy structure and the implications for galaxy formation also relate strongly to the methods by which star formation is triggered in galaxies, as well as the assembly of the supermassive black holes that are found in the centers of galaxies. This also relates to the underlying cosmology, and galaxy structure is likely a new way to determine fundamental features of the universe that can be not be obtained easily through other methods. We describe this work in Chapter 7.

All of this is very exciting for the use of structure with current data, but also with future data such as what will come from the Euclid Space Telescope, the Vera C. Rubin Telescope, James Webb Space Telescope, among others. Now is the time before these telescopes start taking data to discuss and present in a book form the background and the most up-to-date information about galaxy structural evolution and how it can be used to determine fundamental properties of galaxies and the universe itself.

Acknowledgements

I wish to thank my collaborators, students, and post-docs for allowing me to include some of their work in this book, and for the numerous discussions that have significantly increased my understanding of the topic of galaxy formation and galaxy structures. This includes: Asa Bluck, Ken Duncan, Fernando Buitrago, Berta Margalef-Bentabol, Ignacio Trujillo, Will Hartley, Ruth Grützbauch, Sunny Cheng, Leonardo Ferreira, Amy Whitney, Alex Griffiths, and CB Tohill, among others. Support for some of the research presented here came from STFC, NASA, NSF and the Leverhulme Trust and other national and international funding agencies.

Author Biography

Christopher J Conselice

Christopher Conselice is Professor of Extragalactic Astronomy at the University of Manchester, where he has worked since 2020. Conselice is an expert on the properties and evolution of galaxies, and he has investigated these systems from the local universe all the way back to the first galaxies that can currently be observed. Before coming to Manchester, he was a Professor of Astrophysics at the University of Nottingham from 2005 to 2020. Before that he was a National Science Foundation Postdoctoral Fellow at Caltech, and he completed his PhD at the University of Wisconsin-Madison and at the Space Telescope Science Institute in Baltimore, MD.

Christopher won the Leverhulme Prize in 2009 for his research, which is lately focused on the formation of the first galaxies back to start of the universe and preparing for new space missions by NASA and the European Space Agency. He has been on the Thomson-Reuters list of the top 1% most cited researchers in the world for several years. His work has appeared in various media including the Guardian, New York Times, Washington Post, Time Magazine, and many others. In 2016 his press release on discovering that the number of galaxies in the universe was 10 times higher than previous thought—now two trillion—was the 7th most viewed news in all of physical sciences in 2016.

Conselice also sits on many boards and in leadership positions, including as the lead Legacy Scientist for the ESA Euclid mission. He is also the lead editor for the Galaxies and Cosmology corridor of the Astrophysical Journal. In 2018 he was awarded an ERC Advanced Grant to study the first galaxies with James Webb Space Telescope.

Conselice has also published along with historian William Sheehan a popular book on astronomy called "Galactic Encounters" which recounts how we learned about the history of galactic evolution in our universe.

Chapter 1

Introduction

1.1 Introduction to Scientific Classification

When studying any object the most basic thing one can do is determine how that object under examination "looks," and this is often the way in which it is classified and related to other objects. This is certainly the case for classifications of animals and plants in the Linnaean classification of life-forms whereby species identification was, and still is, largely due to the overall appearance of individuals. There are of course objects in which we cannot study their appearances by eye as they are for example very small and not easily seen—good examples of this are atoms and molecular life or DNA. We know that different species of atoms are not classified visually because it is difficult to image atoms and certainly was not possible when the first periodic table of elements was developed by Dimitri Mendeleev in 1869, whereby he classified atomic species by atomic mass and corresponding characteristics of the individual elements.

Similar situations are present to some degree for microscopic objects such as bacteria and cells in biology. Even things we cannot see directly without specialist instrumentation, such as DNA, are best represented as structural objects such as the double helix which after its discovery by Watson and Crick, and others, was awarded the Nobel Prize in 1957. In the case of DNA, the structure of the double helix immediately suggests how it functions, a great example of where the structure of a scientific object immediately suggests its formation, and in this case also its replication method. Such is the ideal situation for classification and function. Finding this type of correlation is not always easy and in the case of galaxies, the focus of this book, it is not always the most obvious features which provide this evidence. We are still largely discovering the best way to both classify galaxies into physically meaningful categories, as well as how to quantify the structural appearances of these systems.

Classification schemes, or in a more general sense the process of taxonomy, has long been a fundamental, and in some ways the most basic activity of science.

doi:10.1088/2514-3433/abb602ch1

The classic taxonomy scheme is the one used to classify life on Earth, and this is also one of the first classification or taxonomy schemes developed by scientists which is still in use today. This was developed by the Swedish scientist Carl Linnaeus in the mid-18th century, who developed a classification scheme that would unite all of the living species on Earth. Within this scheme are a hierarchy of classes: domain (introduced later to include simple forms of life), three kingdoms divided into classes which are divided into orders, and further down into genera which then lead into individual species classification.

Within this Linnaean classification scheme for animals there are various species which are related in their different genius and kingdoms, among others. This classification scheme was originally based almost entirely on how an animal or plant looks. By “looks” we mean how a human observer sees the object with its own eyes. This is also in general how galaxy classification first developed, but it is worth asking if this is the best or only way to carry out such classifications. It is also the case that individuals are within certain species, which are discrete units of classification and not a continuous distribution. Many of the ways in which we approach science, and perhaps society itself, is based on this categorization of things into discrete classes. In many ways astronomy has adopted this method of classifying objects, although for galaxies its usefulness is limited, as we shall discuss later in this book.

This methodology of classification into discrete categories works well in many cases as for the most part individual species of plants and animals are quite distinct from each other. While breeding between two different species is relatively extremely rare, although exceptions exist such as a female horse and a male donkey which can produce a mule, these show that evolution on Earth has produced more or less a large range of quite distinct species. The reason for this is likely an example of survival of the fittest within the theory of evolution, whereby for a species to exist and thrive it must have an evolutionary niche, and this almost always in practice has driven to extinction related species which it has competed with for resources. A good example of this is that there were once on Earth many different species of humans, but they were all driven to extinction (or minor interbreeding with Neanderthals and Denisovans) by the success of homo sapiens, which are the only species of humans alive today. There is, therefore, a reasonable explanation for how we get from taxonomy of classification of species to the underlying scientific reason behind the existence of such a taxonomy.

Taxonomy of living things also provides us with examples of what is not relevant for a classification scheme. Within the classic taxonomy of living things there are distinct categories of species, which although individuals can vary within themselves quite substantially in terms of their weights, heights, coloring, etc. still belong to the same species. For example, a Chihuahua and a Great Dane are very different in appearance and size, but are still of the same species of dog. It is also the case that two members of the same species can be differentiated by these quantities, or by age, but are fundamentally the same. Members of each species are fundamentally extremely similar when compared to each other and when contrasted with individuals from other species. This is another important aspect for classification—sorting out real differences from those which are incidental or due to different ages, as opposed to actual evolutionary differences between species or classes.

There is one exception, even in astronomy, which is how we classify stars. Stars do not have any obvious difference between each other as observed from the Earth, unless very technical and specific observations are made to resolve their disk and structure which we can do for few systems. Astronomers today, as astronomers in the past, differentiate stars visually by their color—that is, how relatively bright a star is in red and blue wavelengths. However, even when we classify stars by their brightness at different wavelengths we still get degeneracies—things look the same but are fundamentally different. An example of this is that blue stars can be either white dwarfs or massive O-type stars which vary between each other enormously in terms of their physical properties and evolutionary history.

It was only after the spectroscopy of stars were acquired in bulk that a suitable classification scheme was developed starting with the Italian Astronomer and priest Angelo Secchi in the 1860s–1870s. This classification scheme for stars was later finally realized to be one of a sequence of temperature, in the familiar OBAFGKM classifications from the Harvard system as discovered by Cecelia Payne in her PhD thesis at Harvard. The remarkable thing about this discovery is that the temperature, and therefore classification sequence, has no obvious pattern within the absorption lines seen in the spectra of stars. However, stars at the same temperature have similar spectra and it was later realized that for main sequence stars temperature correlates with the mass of the stars. This discovery provides a fundamental bedrock for developing theories of stellar evolution—one of the greatest successes of 20th century astronomy.

The similarity between stars and galaxies is however very different as galaxies almost by definition are resolved objects, and therefore they have a structure which can be visually seen by eye in photography or imaging. It is these resolve structures of galaxies and the ways in which astronomers classify and quantify the structures, and the observed evolution of these structures as a function of time which is the focus of this book. The analogy with stars is actually very important in the history of galaxy studies, as it shows that ideas about stars applied to galaxies are more often wrong than right, revealing the danger of extrapolating one system of ideas that is successful for one type of object onto another. It is often illuminating to compare how we think galaxies form compared with how we think stars form. In the field of galaxies initial ideas were often conceived based on our ideas about how stars formed, but were later shown to be incorrect and galaxy structure has been a major way in which we have discovered the likely paths in which galaxy formation progresses. This book explains the various ways in which we have been able to use the resolved structures of galaxies, and how these have changed through cosmic time, to discover how these systems have formed and evolved over 13 billion years of cosmic time.

The resolved structure of galaxies not only gives us a solid way of studying these systems, but these structures are fundamentally the method in which these objects are discovered. This was particularly true in the past when galaxies were first found and cataloged as "nebulae," but it is also true today when there are only a handful of potential galaxies that are relatively unresolved, and what appeared to be point-like objects. These are the quasars which are the cores of more extended galaxies which usually cannot be easily seen behind the bright foreground quasar. Therefore it is the resolved structures of galaxies and what they tell us about the physical state of galaxies which is the focus of this book.

1.2 Historical Background

Speculations on the existence of galaxies really began with ideas for what the Milky Way was. The Milky Way in this sense means the observational phenomenon of a band of light that crosses the sky at certain locations and times of the year, nominally in the summer time in the northern hemisphere. There have been many speculations about the origin of this diffuse light that originated in cultures throughout the world. The idea that this light is in fact produced by unresolved stars was an idea of the ancient Greeks, starting with Democritus who is better known today as one of the originators of the atomic hypothesis.

From the vantage point of a dark location on Earth there are a few extragalactic objects that can be seen by eye, including the Magellanic Clouds, Andromeda and a few others. It is likely that anyone in early history considering the origin of these diffuse systems thought of them as related to the Milky Way light, as by eye they appear to have the same diffuse nature. In a sense this is true as extragalactic systems appear diffuse because we are seeing unresolved stars, just as when we examine the Milky Way. Of course, the vast distances and the inherent uniqueness of these objects that we would call external galaxies today, from the diffuse light of Milky Way was not, and could not, of been appreciated in ancient times.

Perhaps remarkably there does not seem to be any mention of nebula we call galaxies today, such as Andromeda, in the “Almagest,” Ptolmey’s book on star positions written in 125 AD. It was not until the book “The Book of the Fix Stars” written around 964 AD by Abdul-Rahman al-Sufi, that a scientist considered that Andromeda could be a separate object to the Milky Way, although no evidence of that could be had at that time. Very little could be done in terms of studying these objects until the invention and use of the telescope in 1609–1610.

Soon after the telescope was constructed and pointed at the sky, what were called “nebula” were found and noted by different observers. This includes many objects that would later become part of the first catalogs of nebula, which of course included galaxies. It is important to realize that by “nebula” these early observers referred to any object that was resolved and looked “diffuse”—not a point source such as a star. Today we know that these were not only galaxies, but also consisted of star clusters and gaseous nebula within our own Galaxy. Nearby star clusters and distant galaxies look indistinguishable when viewed by eye on an optical telescope.

In many ways, the systematic study of galaxies began with the Messier catalog constructed by Charles Messier, who observed and discovered these objects at the Hôtel de Cluny on the left bank of Paris. Through visual sweeps of the sky with a telescope, Messier compiled a list of 110 nebulous non-cometary objects—essentially objects that did not move. Messier was interested in finding comets, and essentially had no interest in the nebula themselves. He is vastly more famous for his list of nebula than any of the comets he found, but he created this list as a guide for observers for what to discount as possible new comets. Interestingly, a large fraction of the nebula in Messier's list were not found by him, but by other astronomers before him, many at the Paris Observatory just south of where he worked.

The list of known nebula was later expanded by the more systematic catalogs of John and William Herschel in the southern and northern hemispheres. This work really began the first serious and systematic study of nebulae. The Herschels' used increasingly larger telescopes starting in the late 18th century to catalog the location of nebular objects by using what are called "star sweeps." Within these star sweeps where an observer would look through a telescope as it moved across the sky and any objects that showed a nebulous appearance, basically diffuse light, would be cataloged in what became the "Catalogue of Nebulae and Clusters of Stars" (e.g., Herschel 1786). This list of nebula created over decades by the Herschels ultimately numbering over 5000 objects became the "general catalog" of nebula, Most of the bright nearby galaxies in which Messier and the Herschels classified as "nebula" were actually what we would today call galaxies, but not all. Many of them are globular clusters, reflection nebula, and gaseous nebula within our own Galaxy, as described earlier. However, most of them are extragalactic, especially the fainter ones. This General Catalog was later updated by the Danish astronomer John Louis Emil Dreyer who created the New General Catalog with 7840 entries from which the very common "NGC" numbers arise, which is by far the most commonly used catalog of galaxies today for nearby systems.

Finding galaxies through their "nebulosity" is one thing, but understanding what the nebula were and realizing that some of these are Galactic—within our own Galaxy—while others are extragalactic—outside the Galaxy, took some time to develop. First, it was important to get some idea of the structures of these systems. Visually based drawings of galaxies provided the first morphological studies of these systems. When did galaxy structure actually start being noted and used? This was first carried out by the Herschels, Lord Rosse in Ireland, and William Lassell in England.

Very early on, the Herschels would make notes of unusual features of their nebula, as they were able to do more than just identify these, but made comments on their appearance and used descriptions such as "round," "extended," "irregularly," and even "pretty" (see Figure 1.1). Later, astronomers with even larger aperture telescopes were able to discover the spiral pattern of galaxies within some of these systems. Most famously this was carried out by William Parsons, 3rd Earl of Rosse in Ireland who examined, among other galaxies, the Whirlpool nebula, what is often called M51, as far back as 1845 (see Figure 1.2). While Rosse's work is by far the most famous today from this era, the claim was made by astronomers in the 19th century that Lassell's drawings were the most accurate (e.g., Roberts 1889). Quite clearly visual examination of these systems provided the first evidence that galaxies had a real global structure that differed from one object to the next. The physical origin of such a structure was still unknown however.

Very soon after Lord Rosse's and other astronomer's drawings of nebulae were made, photography was used to study galaxies, which allowed a permanent largely unbiased view of galaxy structure. Photography remained the standard way to study

The abbreviations are to be underſtood as follows.

B. Bright.	v. very.
F. Faint.	c. conſiderably.
L. Large.	p. pretty.
S. Small.	e. extremely.

Of theſe letters I have compoſed vB. cB. pB. pF. vF. eF. vL. pL. pS. vS. eS.; all which require no farther explanation.

R. Round.	l. a little.
E. Extended.	i. irregularly.
M. in the middle.	g. gradually.
b. brighter.	ſ. ſuddenly.
m. much.	

Figure 1.1. Descriptions of terms used to describe the nebula discovered by William Herschel in his 1786 paper "Catalogue of One Thousand new Nebulae and Clusters of Stars." The often used term "irregular" to describe galaxies has been in use since this very earliest published catalog. Note the use of the long S in this text for the word "suddenly."

Figure 1.2. Sketch of M51 by the Lord Rosse as taken from his observatory at Birr Castle in Ireland in 1845. This is one of the images that began the study of galaxy structure and morphology. A modern version of this galaxy taken with deep imaging on telescopes is discussed in Chapter 8.

galaxy structure from the mid-late 19th century up until the 1980s when digital detectors took over. It is interesting to note that the photographic work done in the late 19th century was, even by modern standards, fairly impressive with many of the features of well known galaxies already quite easily visible.

It was not until photography was applied to astronomy that the true morphological and structural nature of the nebulae were revealed in enough detail for many galaxies such that real progress could be made in galaxy structure studies. One of the pioneers of this work was the Welsh astronomer Isaac Roberts who used his observatory in Sussex in the late 19th century to take images of famous galaxies, such as M51 and Andromeda, which appear very similar to images taken today (see Figures 1.3 and 1.4). It was clear in these images that these nebula were spiral in shape, which most nearby bright galaxies turned out to be (Figure 1.4). However, Roberts and other astronomers at the time also took deep photographs of Galactic nebula, such as the Orion Nebula, M42, as well as comets and other faint astronomical objects. These pictures themselves did not give astronomers an unambiguous idea that these objects were outside our own galaxy, but they were able to start showing their true nature, and the resolved structures of these galaxies was the key aspect to this. At a similar time, deep spectroscopy of these objects showed that the spectra of at least the brightest portions of Andromeda have similar features to stellar spectra. However, it was not until the 1920s when astronomers generally accepted that a fraction of the nebulae were outside our own Galaxy.

At the time it was thought by some that these features—the spiral arms and bright knots in the arms of these nearby galaxies were proof of the nebula hypothesis,

Figure 1.3. One of the first photographic images of the Andromeda galaxy, M31, as imaged by Isaac Roberts in 1888 using his 20 inch Grubb telescope from his observatory in Sussex, England.

Figure 1.4. Images of nearby galaxies as taken from the Hubble Space Telescope. Roughly shown on the top are ellipticals and S0s, while the middle row are early-type spirals with larger bulges, and the bottom row are systems with a larger disk to bulge ratio—the late-type spirals. Courtesy of NASA.

which was the idea that these systems were at the distances of stars in our own Galaxy, and are forming into solar systems like our own (e.g., Moulton 1905). This was a remarkable theory and it is incredible today to examine the photos astronomers at the time were able to acquire and compare them favorably with ones we can obtain today. Even with these photos these astronomers developed utterly different ideas about what the structures of these systems implied. In this sense, the structure of the nebula or what we call galaxies were misleading to many astronomers as the interpretation of the structure was incorrect. This is why it is important to have some physical understanding of systems and what the structures imply before using these features to derive the nature of the objects. Perhaps another lesson to learn from this, from an astronomical perspective, is to not interpret a structure until the distance is known.

The nebular hypothesis was first proposed by the French mathematician and astronomer Pierre-Simon Laplace who came up with an early idea for explaining how planets in our solar system formed. There is nothing wrong inherently with the nebular hypothesis, it even remains a valid hypothesis today for explaining how planets and solar systems formed. The basic idea is that the star forms first and surrounding the star is a large rotating disk of gas that over time will condense into planets. When spiral galaxies were first imaged, it appeared to astronomers at the time perfect evidence for solar systems in formation. The Chamberlin–Moulton hypothesis further speculated that the spiral nebulae were these systems in formation, which were triggered by the gravitational pull from a nearby star passing by. The debris pulled out would then condense in this theory into orbiting planets. A modern interesting side note to this is that spiral structures are in fact seen around forming stars that are likely forming into planets as seen with modern sub-mm telescopes, providing some evidence for the formation hypothesis of Laplace, although on a much smaller physical scale that would have been implied by the angular sizes of galaxies.

At this point in the history of early galaxy studies, there were many images of different systems and it was clear that galaxies can look very different from one another based on photographs of galaxies which increased to around 1000 systems by 1918 (Curtis 1918). Naturally, astronomers started to think about ways to link different systems and the way to do that is to develop a classification scheme. Early notable systems for classifying galaxies into a consistent system were developed by Wolf (1908) and Lundmark (1926), among others. This work was superseded by what we call today the Hubble system, which was published in its modern form in Hubble (1926), with the famous "Hubble Tuning Fork" established in Hubble (1936) and Sandage (1961). The Hubble sequence (Figure 1.5) has two main types of galaxies, ellipticals, and spirals, with a further division of spirals into those with bars and those without bars. Hubble, and the astronomers who followed him, could classify most nearby bright galaxies in terms of this system of classification.

The development of morphological classification methods continued after Hubble's work, mostly building upon his ideas. Most of the newer systems developed in the 20th century were based for the most part on visual impressions without any regard for the underlying causes of those visual effects. One famous case is the system developed by de Vaucouleurs (1959) who created a revised version of the Hubble sequence, which took into account properties such as bars, rings, and

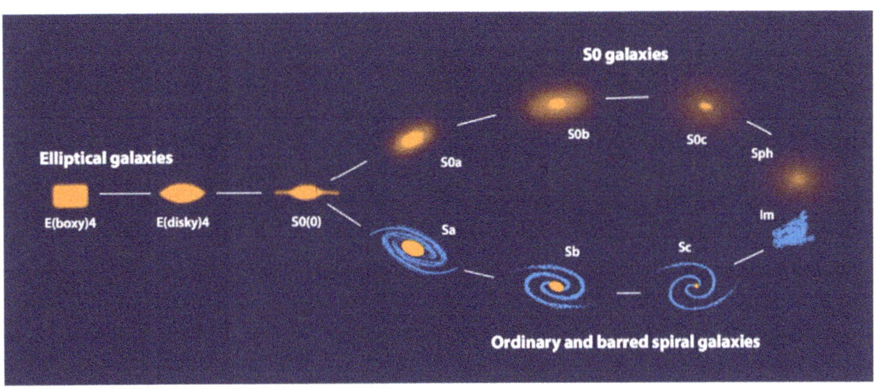

Figure 1.5. A modern view of the Hubble sequence. This is the traditional view of galaxy morphology, such that ellipticals are on the left hand side and the spiral galaxies are on the right. The spirals are further divided into different types depending on whether they have a bar in their center or not. Reproduced from Kormendy & Bender (2012). © 2012. The American Astronomical Society. All rights reserved.

other internal features that were easily seen on photographic plates of galaxies. Soon after that van den Bergh (1960, 1976), and later Elmegreen & Elmegreen (1987) developed systems of classification that were based on the form of spiral arms, and the apparent clumpiness of the light in these arms.

The visual classification of galaxies is important, as we shall see, and all systems have some use. However, all features should be explained by physics and it remains true that not all features that can be seen by eye are important for understanding the physical and evolutionary state of galaxies. Just by a casual inspection it is not obvious which structural features of galaxies are fundamental to their formation history. Any morphological or structural system needs to be shown to be useful for understanding galaxies, and not just for describing their most obvious features in visual light photographs. An example of this would be to try to classify animals by hair or eye color—neither of which are fundamentally important for the type of animal they are. As there is now an extensive amount of photometric and spectroscopic data we can determine what are the most fundamental measures of stellar populations and dust/gas properties in galaxies, which we discuss later in this book.

At about the same time as these early morphology systems were developed progressively more was learned about galaxies and their properties beyond structure. This includes works by astronomers such as Holmberg (1958) who established that physical properties of nearby galaxies correlate with the structure of a galaxy, namely its Hubble Type. Holmberg (1958) furthermore showed that ellipticals are usually massive and red with at most small amounts of star formation. On the other hand, spirals are less massive, bluer and are undergoing a significant amount of star formation in comparison to the existing amount of stellar mass. These trends also hold when examining other physical parameters as well (e.g., Roberts 1963; Roberts & Haynes 1994; Conselice 2006; Allen et al. 2006). In general, early types are more likely to be evolved galaxies with no gas or ongoing star formation, while later type galaxies are more likely to have existing star formation. As we will discuss in later chapters, the segregation of galaxy properties into morphology in the local universe provides an important clue for understanding the physics of galaxy formation. One way to see this is that for nearby galaxies the environment strongly correlates with a galaxy's morphology (e.g., Dressler 1984; see Chapter 6).

The next major advances in morphological and structural examinations of galaxies came about with the advent of photometric photometry, this was especially true with the advent of charged coupled devices (CCDs) in the 1980s, which allowed for linear and quantitative measurements of light distributions in galaxies. One of the first major contributions using this technology was done by de Vaucouleurs (1948) who showed by using photometry that the light profiles of ellipticals can nearly all be fit over their entire structure by a universal light distribution form that is now known as the de Vaucouleurs profile, which is discussed in detail in Chapter 3.

These types of analyses of the one-dimensional light profile of galaxies were expanded by others, most notably Sérsic (1963). Sérsic demonstrated that all galaxies, no matter their overall morphological type, could be well fit by a generalized light distribution. What is found is that, in general, lower mass galaxies and disks contain light profiles that are exponentials, while massive ellipticals generally following the de Vaucouleurs profile. This has led to a huge industry in measuring the light profiles of galaxies in the nearby and distant universe, which continues through to today (see Chapter 3).

From the 1970s onward the structural study of galaxy structure started to become more complicated with the fitting of light profiles including multiple components as well as extra features beyond the gross light profile, such as bars and other structures. The most famous and common use of this is the decomposition of galaxy light into bulge and disk profiles (e.g., Kormendy 1977). It is also now possible to fit features such as bars, rings, and lenses when decomposing a galaxy's light distribution (e.g., Kormendy 1979; de Vaucouleurs et al. 1991).

Another major question is how to understand and recover the three-dimensional structure of disk galaxies. This was investigated in several studies (e.g., van der Kruit & Searle 1982), which included detailed examination of the shapes of bulges and disk in spiral systems (e.g., de Jong 1996; Peletier & Balcells 1996). Part of this work naturally has led to trying to understand the origin of these structures.

Although a large amount of work has been done on examining the structures and morphologies of galaxies in the nearby universe it is a challenge to probe these features when studying more distant galaxies. This is because current technology does not permit distant galaxies to be resolved in the same detail as we can for more nearby systems. One result of this is we can make progress in studying galaxy structure and morphology when we can accurately measure properties of distant galaxies, and compare with nearby galaxies to determine evolution. This is one of the main goals of this book. Although there are many methods for studying galaxy evolution this book argues and shows that structural analyses are now an essential aspect for studying these processes, and can even provide unique information on the history and physics of galaxy assembly.

1.3 Stranger Things—Irregular, Peculiar, and Dwarf Galaxies

Galaxy studies typically focus on normal systems, with unusual ones considered rare and often simply curiosities that can be ignored. After galaxies were first studied in the 18th and 19th centuries very few of the ones examined look like what we would today call peculiars or irregulars. In fact, these systems were not accounted for at all within Hubble's classification system, nor in other competing ones. However, what we now know is that these "unusual" galaxies are actually the ones most important for understanding the assembly of galaxies, as they provide rare stages of transformation. In the case of dwarf galaxies, these are the most common type of galaxy in the universe, yet they are extremely difficult to study and measure due to their low surface brightness and faint magnitudes.

So-called peculiar features in both galaxy structure and spectra have in the past provided major clues for the formation methods of galaxies. Therefore investigating unusual or strange features in galaxy structure and spectra have often driven the discovery of new astronomical objects such as active galactic nuclei (AGN) and eventually central massive black holes. Likewise, galaxy structural peculiarities hold important clues to the formation of galaxies as these features are often evidence that

Figure 1.6. Optical images of galaxies taken by the Hubble Space Telescope in various stages of merging in the nearby universe. Some of these mergers are just starting and can be characterized as galaxy interactions. Other systems are in various post-merger phases where two galaxies have merged to become a single system. Some of these galaxies were originally classified by Arp (1966). Courtesy of NASA.

galaxy mergers are a process responsible for some aspects of galaxy formation and evolution (Figure 1.6).

Some examples of peculiar features in galaxies, in this case, unresolved imaging and spectra, that turned out to be important include quasars, which appeared as galaxies with very bright nuclei (e.g., Schmidt 1963) that sometimes also contain bright radio emission. Another related object discovered by serendipity are the active galactic nuclei, galaxies which were discovered to have peculiar spectra of otherwise normal looking galaxies (Seyfert 1943). While it is true that galaxies with these properties are rare, their unique observational features can tell us about relatively quick phases of galaxy formation and evolution. The same is true for peculiar galaxies which occur when galaxies are assembling at a fast rate typically due to the merger process.

A good analogy of this is how humans and other animals are able to grow—which is from nutrition by eating. If we take humans for example, most of us only eat for a short period of our day (as opposed to some animals like Pandas that eat continuously when awake). If we knew nothing about humans and simply took pictures of them at random times, we would rarely see someone eating—if we consider that a typical human spends two hours a day eating, then we would only see someone eating ~8% of the time. One could conclude based on the preponderance of pictures of humans not eating that the process is not important, but of course that would be incorrect. Likewise, we may see peculiars and AGN features for only a short period of a galaxy's life-time, but these are the critical stages when central massive black holes and the galaxy itself are assembled.

Of course, we also need to know the duty cycle and history of these processes in galaxies to fully understand their role in forming galaxies and their components. This timescale issue is a very important one in astronomy and is something that we will discuss in more detail when we describe how to understand the role of peculiar galaxies, which are often the merger of two separate galaxies forming into a new system.

After photography was used to image galaxies in the early 20th century some peculiar galaxies were discovered but they were usually just considered oddities. The majority of galaxies imaged with early photography turned out to be spiral galaxies, the most numerous type of bright galaxy in the nearby universe. The first galaxy classification schemes developed by, e.g., Wolf (1908), Lundmark (1926), and Hubble (1926) typically did not include unusual or peculiar galaxies in their systems. This is no doubt due to the fact that nearby peculiar galaxies are very rare, accounting for about 1% of the nearby population and tend to be among the fainter lower mass systems.

Deeper photography in the 20th century revealed more representative galaxies showing that there were systems that are peculiar and irregular in appearance and could not be classified into any of the popular classification schemes. Astronomers who studied these, including Arp (1966) concluded that these galaxies indeed had a very unusual origin, which was not initially thought to be due to merging. Some thought that these galaxies were in some form of early-formation, or that we are witnessing the formation of galaxies driven by magnetic fields. Perhaps surprisingly, it was not generally considered that these structures were dominated by gravitational attraction combining two galaxies into one. Later it was shown that many of these peculiars are undergoing high star formation rates and many of them are Ultra Luminous Infrared Galaxies (ULIRGs).

Because they are relatively rare for bright galaxies it might seem surprising that peculiar galaxies hold important clues for the formation of all galaxies. It is largely the case that at one time or another nearly all galaxies, or their ancestors, would have been classified as a peculiar. It is certainly the case that when we examine the morphological evolution of galaxies, by looking at distant systems, we see that peculiars were much more common in the past (e.g., Mortlock et al. 2013). Even though the Hubble (1926) tuning fork diagram has no location to classify peculiar galaxies eventually its expansions did include unusual galaxies.

In the 1970s astronomers began to recognize that these peculiar galaxies are indeed probably the result of merging systems. We now understand through discussions in Chapter 4 that galaxy mergers and interactions are one of the dominant ways in which galaxies form. Mergers are also responsible for inducing formation processes for stars, planets, and black holes during cosmic time, which will be discussed in Chapter 6. Overall we are just now learning about the full observational role of galaxy mergers, and how they influence the formation of galaxies themselves.

It is important to realize that irregular galaxies and peculiar galaxies differ, even though their names are similar. Astronomically speaking an irregular is a lower mass dwarf galaxies whose structure is dominated by star formation while the peculiars, as described above, are those undergoing merging events. These irregulars have a large star formation which typically is not triggered by galaxy interactions or mergers, but by dynamical quiescent star formation.

Another type of galaxy are the dwarf elliptical galaxies, which dominate the galaxy population in terms of number and are also among the lowest surface brightness galaxies in the universe (e.g., Conselice et al. 2001). Galaxy clusters are well known for hosting massive galaxies, but beginning with Reaves (1956) astronomers realized that the most common galaxy type in clusters are these small dwarfs (e.g., Marzke & da Costa 1997). After a short time these dwarfs should be destroyed and become part of the intracluster medium of the cluster (Gibson et al. 1997). Their presence has therefore long been a mystery (e.g., Ferguson & Binggeli 1994) given their apparent undisturbed morphologies unless these systems are dominated by dark matter (Penny et al. 2009). Dwarf ellipticals in clusters differ from dwarf ellipticals in the local group in several important ways (e.g., Ferguson & Binggeli 1994) and may turn out to be relatively unevolved primordial galaxies with large amounts of dark matter. Figure 1.7 shows images of these dwarf galaxies, including dwarf ellipticals and dwarf irregulars.

Figure 1.7. Images of dwarf irregulars and dwarf ellipticals in the nearby universe. These galaxies are very low surface brightness, which makes them difficult to study and discover. Courtesy of NASA.

1.4 Low Surface Brightness Galaxies

One of the newest and least explored areas of galaxies is the exploration of low surface brightness (LSB) light and LSB features of galaxies. We already discussed the dwarf galaxies above, which are often of extremely LSB and were not discovered for several decades after the first galaxies were identified. There are other LSB galaxies that are still being characterized and studied, as well as searches for outer diffuse light in more massive galaxies. Typically these systems are larger and more massive than the more common dwarf ellipticals.

Many galaxy catalogs contain objects that have been obtained from magnitude or diameter limited surveys. Sources with surface brightness near or below that of the sky background are therefore underrepresented in such surveys. A class of galaxies known as LSB galaxies, with central disk surface brightness μ_0^B a few magnitudes greater than the Freeman (1970) value of $\approx$ 21.65 mag arcsec^{-2}, have been detected and extensively studied (see review by Impey & Bothun 1997). The space density of these objects is poorly known, but statistical arguments indicate that 30%–50% of the total galaxy population may consist of LSB galaxies (de Blok 1997), but ultimately understanding this is still very uncertain. The properties and evolution of LSB galaxies are therefore essential in constructing a complete picture of galaxy evolution.

Some ideas on the origin of large LSBs includes that they are systems in an early stage of disk formation. The ISM of these objects is metal-poor and perhaps dynamically hot, which prevents cooling and limits star formation. The disks of LSBs are large with low surface mass densities, and by being embedded in massive haloes this stabilizes the disks and prevents significant star formation. In this picture, LSB galaxies thus appear to be very slowly evolving systems, free from instabilities and interactions common among high surface brightness (HSB) galaxies. As such, they provide valuable insight into the nature of galaxy formation and evolution including their structures and morphologies.

1.5 Primer on Galaxy Structure and Galaxy Formation

A book like this must necessarily discuss other properties of galaxies, especially those which relate to structure. This includes topics such as star formation, black holes, as well as the processes which drive the formation of galaxies to begin with. Mergers will be a major aspect of our discussion, but other processes, such as gas accretion and feedback (any process that goes against the pull of gravity) can also be related to the structures of galaxies.

Below we give a broad overview of the modern view of how galaxy formation and evolution progresses. One way to start this discussion is to consider the final product (so-far) of these processes—the local universe. We now have a fairly solid understanding of the galaxy population in the local universe based on large surveys such as the Sloan Digital Sky Survey (SDSS) and GAMA (e.g., Conselice 2006; Driver et al. 2011). What we find within nearby galaxies is that ellipticals and spirals with active star formation dominate the local universe in terms of the stellar mass content. Galaxies which are higher mass are also more likely to be elliptical and the

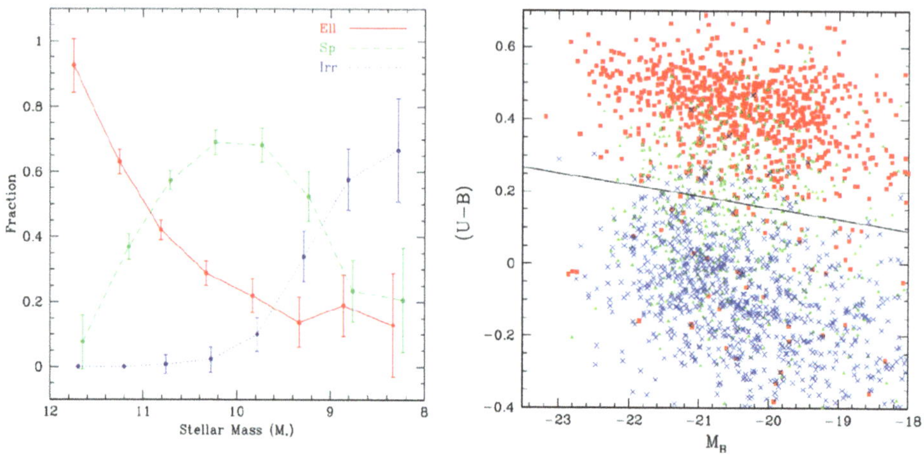

Figure 1.8. Statistical view of nearby galaxies from Conselice (2006). The left hand side shows the distribution of different galaxy morphologies as a function of stellar mass. This shows that the more massive a galaxy is, the more likely it is to be an early type galaxy or elliptical. The right hand part of the figure shows the color–magnitude diagram for galaxies labeled by morphology, such that the red points are early type ellipticals and the blue points are the spiral/disk galaxies. As can be seen, the ellipticals are predominately red and the disk galaxies are blue.

lowest mass galaxies are more likely to be irregular in morphology (Figure 1.8). However, it is not the case that spirals and ellipticals within the Hubble sequence are the most common or even "typical" galaxies. As mentioned, low mass galaxies, or dwarfs, dominate in number and are the most common (e.g., Conselice et al. 2001). We can measure the stellar and luminosity functions of nearby galaxies to a high accuracy (e.g., Loveday et al. 2012), the result of this is often used to calibrate models of galaxy formation. We can also measure when the stars in these galaxies formed in the past based on detailed stellar population analyses using spectra or spectral energy distributions from photometry.

We are however still trying to understand how and when these galaxies assembled. There are in general two ways in which galaxies can form and evolve —from internal processes, which are sometimes called "secular" and from external processes such as gas accretion and mergers with other galaxies. There have been several investigations that claim that it is secular evolution within disks that can explain how their internal bars, rings, and lenses form (e.g., Kormendy 1979; Combes & Sanders 1981). Secular effects, not driven by hierarchical assembly, are also perhaps responsible for the formation of so-called "pseudo-bulges" which are perhaps related to the formation of central massive black holes (e.g., Kormendy & Kennicutt 2004; Sellwood 2013).

We have made some progress on addressing the issues related to this in the last 10–15 years. As an example, the first measurements of galaxy properties at high redshift showed that the star formation rate increases with look back time (e.g., Madau et al. 1998), and peak at redshift $z \sim 2$ and then declines again at $z > 2$ (Madau & Dickinson 2014). These types of observations, however, mostly reveal when stars form, but not how galaxies assemble.

This relates to the underlying cosmological model of galaxy formation, that due to observations on large-scales—including the cosmic microwave background, we have good reasons for assuming approaches the correct model. This is the ΛCDM model—whereby we have a universe dominated by dark energy, as given by the Λ term, and cold dark matter (CDM). A major prediction of this model is that galaxies should form hierarchically, that is that massive galaxies should build up by merging with and accreting other galaxies. This is directly a result of the dark matter particle being cold, that is slow moving. A slow moving particle will gravitationally collapse on scales that are small, as the streaming distance for such a slow moving particle will be short before gravitational collapse occurs, giving a large number of small objects. Furthermore, within this picture, these first forming objects are not galaxies, but the dark matter halos in which galaxies later form out of. At the same time that these early dark matter halos collapse, so does primordial gas, which at this point mostly consists of hydrogen and helium. This gas is also hot, in fact it is too hot to form stars. The further formation of galaxies will depend on the cooling of this gas, the accumulation of new gas, and the mergers of the galaxies themselves. Galaxy structure can provide clues for all of these processes to one degree or another.

As we will discuss in detail in this book, we know that mergers are occurring based on observing them throughout the history of the universe (e.g., Le Fèvre et al. 2000; Conselice et al. 2003, 2008; Hammer et al. 2009; Lotz et al. 2008; Bluck et al. 2009, 2012; Man et al. 2012; López-Sanjuan et al. 2012; Mundy et al. 2017; Duncan et al. 2019). As we will see, we can also view the star formation in galaxies through structure, as often this shows up and is revealed through small-scale features in galaxies, which are created from the clusters of stars being formed. Other modes of formation are also seen and can be measured, such as gas accretion, which while not directly seen through structure can be inferred (e.g., Conselice et al. 2013). That is, stars can form in multiple galaxies and no further star formation can occur but galaxy assembly can progress through the mergers of these individual systems.

Observationally, the approach that is quite often taken toward understanding the physical origins of galaxies is through galaxy observables, such as luminosity/mass functions and scaling relations such as internal velocities versus luminosities to test models. These models, starting with simple collapse ones (Eggen et al. 1962) implement basic physics, and the idea is that these models are subsequently able to predict what we see in the observations. Application of theory in this way has revealed many clues toward understanding how galaxy formation has occurred, although as we discuss in Chapter 7 there are still significant problems with theory matching galaxy observables. However, this is a good thing as it means that the problem of galaxy formation and evolution is not completely solved yet and there are other issues that need to be addressed before we can claim victory over this topic.

1.6 Book Structure

In this book an up to date summary of what we know about galaxy structure throughout cosmic time is described. This includes methods for how we measure structure and the various ways it can be quantified. We also discuss ways in which structure can tell us about the formation and evolutionary history of galaxies, including the formation of stars and their central massive black holes. Furthermore, structural evolution is a new and exciting way to test theoretical models of the formation of galaxies, as well as a new approach toward understanding the cosmological parameters of the universe including the nature of dark matter. We finish by talking about novel aspects and the future of galaxy structure using new data, surveys, and techniques.

This book is structured as follows. Chapter 2 provides a philosophical basis for the book and takes a scientific view of whether galaxy structure and morphology is something that is worth measuring and how we might go about doing this. Chapter 3 is a detailed chapter on the ways in which galaxy structure is quantified and measured. This includes understanding the limitations of how we can measure structure and in what situations we can quantify it. Chapter 4 introduces the distant universe and the basic observations concerning what we see when we image galaxies that existed in the distant past. Chapter 5 expands on this and discusses how we can quantify the ways in which galaxies have changed in shape and structure over 13.5 billion years of cosmic time. Chapter 6 is about the use of galaxy structure as a way to determine how galaxy structure relates to star formation triggering as well as supermassive black hole build up. Chapter 7 describes how structure can be used as a tool to uncover other aspects of astrophysics, including cosmological measures. Chapter 8 is the concluding chapter that goes into some detail about future uses of structure and morphology, including new facilities such as the James Webb Space Telescope, Euclid, and Rubin telescopes, non-optical morphologies, high-resolution imaging as well as how to apply machine and deep learning to answer outstanding problems and questions.

Whenever observational measurements are quoted and discussed we use a cosmology such that $\Omega_\lambda = 0.7$ and $\Omega_m = 0.3$ with $H_0 = 70$ km s^{-1} Mpc^{-3}. The exception to this is when we discuss how the merger history of galaxies can be used a probe of these cosmological parameters and place constraints on them.

References

Allen, P. D., Driver, S. P., Graham, A. W., et al. 2006, MNRAS, 371, 2

Arp, H. 1966, ApJS, 14, 1

Bluck, A. F. L., Conselice, C. J., Bouwens, R. J., et al. 2009, MNRAS, 394, L51

Bluck, A. F. L., Conselice, C. J., Buitrago, F., et al. 2012, ApJ, 747, 34

Combes, F., & Sanders, R. H. 1981, A&A, 96, 164

Conselice, C. J., Gallagher, J. S., & Wyse, R. F. G. 2001, ApJ, 559, 791

Conselice, C. J., Bershady, M. A., Dickinson, M., & Papovich, C. 2003, AJ, 126, 1183

Conselice, C. J. 2006, MNRAS, 373, 1389

Conselice, C. J., Rajgor, S., & Myers, R. 2008, MNRAS, 386, 909

Conselice, C. J., Mortlock, A., Bluck, A. F. L., Grützbauch, R., & Duncan, K. 2013, MNRAS, 430, 1051

Curtis, H. D. 1918, Publications of Lick Observatory, 13, 9

de Blok, W. J. G., & McGaugh, S. S. 1997, MNRAS, 290, 533

de Jong, R. 1996, A&A, 313, 45
de Vaucouleurs, G. 1948, AnAp, 11, 247
de Vaucouleurs, G. 1959, Handbuch der Physik, 53, 275
de Vaucouleurs, G., de Vaucouleurs, A., Corwin, H. G., et al. 1991, Third Reference Catalogue of Bright Galaxies (New York: Springer)
Dressler, A. 1984, ARA&A, 22, 185
Driver, S. P., Hill, D. T., Kelvin, L. S., et al. 2011, MNRAS, 413, 971
Duncan, K., Conselice, C. J., Mundy, C., et al. 2019, ApJ, 876, 110
Eggen, O. J., Lynden-Bell, D., & Sandage, A. R. 1962, ApJ, 136, 748
Elmegreen, D. M., & Elmegreen, B. G. 1987, ApJ, 314, 3
Ferguson, H. C., & Binggeli, B. 1994, A&ARv, 6, 67
Freeman, K. C. 1970, ApJ, 160, 811
Gibson, B. K., Loewenstein, M., & Mushotzky, R. F. 1997, MNRAS, 290, 623
Hammer, F., Flores, H., Puech, M., et al. 2009, A&A, 507, 1313
Herschel, W. 1786, Philosophical Transactions of the Royal Society of London Series I, 76, 457
Holmberg, E. 1958, Meddelanden fran Lunds Astronomiska Observatorium Serie II, 136, 1
Hubble, E. P. 1926, ApJ, 64, 321
Hubble, E. P. 1936, Realm of the Nebulae (New Haven, CT: Yale University Press)
Impey, C., & Bothun, G. 1997, ARA&A, 35, 267
Kormendy, J. 1977, ApJ, 217, 406
Kormendy, J. 1979, ApJ, 227, 714
Kormendy, J., & Kennicutt, R. C. 2004, ARA&A, 42, 603
Kormendy, J., & Bender, R. 2012, ApJS, 198, 2
Le Fèvre, O., Abraham, R., Lilly, S. J., et al. 2000, MNRAS, 311, 565
López-Sanjuan, C., Le Fèvre, O., Ilbert, O., et al. 2012, A&A, 548, A7
Lotz, J. M., Davis, M., Faber, S. M., et al. 2008, ApJ, 672, 177
Loveday, J., Norberg, P., Baldry, I. K., et al. 2012, MNRAS, 420, 1239
Lundmark, K. 1926, Arkiv for Matematik, Astronomi och Fysik, 19, 1
Madau, P., Pozzetti, L., & Dickinson, M. 1998, ApJ, 498, 106
Madau, P., & Dickinson, M. 2014, ARA&A, 52, 415
Man, A. W. S., Toft, S., Zirm, A. W., Wuyts, S., & van der Wel, A. 2012, ApJ, 744, 85
Marzke, R. O., & da Costa, L. N. 1997, AJ, 113, 185
Moulton, F. R. 1905, ApJ, 22, 165
Mortlock, A., Conselice, C. J., Hartley, W. G., et al. 2013, MNRAS, 433, 1185
Mundy, C. J., Conselice, C. J., Duncan, K. J., et al. 2017, MNRAS, 470, 3507
Peletier, R. F., & Balcells, M. 1996, AJ, 111, 2238
Penny, S. J., Conselice, C. J., de Rijcke, S., et al. 2009, MNRAS, 393, 1054
Reaves, G. 1956, AJ, 61, 69
Roberts, I. 1889, MNRAS, 49, 389
Roberts, M. 1963, ARA&A, 1, 149
Roberts, M., & Haynes, M. 1994, ARA&A, 32, 115
Sandage, A. 1961, The Hubble Atlas of Galaxies (Washington, DC: Carnegie Institution)
Sellwood, J. A. 2013, in Planets, Stars and Stellar Systems, Vol 5 (Berlin: Springer), 923
Sérsic, J. L. 1963, BAAA, 6, 41
Seyfert, C. K. 1943, ApJ, 97, 28
Schmidt, M. 1963, Natur, 197, 1040

van den Bergh, S. 1960, ApJ, 131, 215
van den Bergh, S. 1976, ApJ, 206, 883
van der Kruit, P. C., & Searle, L. 1982, A&A, 110, 61
Wolf, M. 1908, Publikationen des Astrophysikalischen Instituts Koenigstuhl-Heidelberg, 3, 109

Chapter 2

Galaxy Structure and Its Meaning

In this chapter, a presentation of the underlying meaning for morphology and structures of galaxies is discussed. This is an important aspect as we will need to understand the importance of galaxy classification, morphology, and structure and to justify its existence as a scientific methodology before we can take confidence in the results that are obtained from it. This is in many ways the philosophical basis for much of the rest of the book.

It is important to ask if we even need a physical classification system for galaxies, and does it make sense to even have one? It might be the case that the Hubble sequence, and all later morphological systems, are a distraction from more physical properties of galaxies. In this sense, galaxy structure would not be fundamental—it is simply just the "weather" in a galaxy—a temporary state that is not fundamentally related to the underlying galaxies themselves. In a real sense, the galaxy itself—that is the visible light we see from stars—is in fact like weather as the vast majority of the mass of a galaxy is composed of dark matter. We address this point later in Chapter 8, but suffice it to say it does appear that the stellar light distribution of galaxies follows in many ways the dark matter distribution.

Because of this, it is important to ask if it is possible that relations between observables and non-morphological galaxy properties are sufficient, and superior to structure, to characterize the population. As an example, we have properties of the underlying kinematics of galaxies, the velocity of rotation, and the internal velocity dispersion. Could this combined with the stellar mass or absolute magnitude be superior or replace galaxy structure? Also, if we are to classify and measure the structures of galaxies we want to understand what it tells us and how we can use that information to derive the physical state of a galaxy, and just as importantly how it formed and evolved.

As discussed in the first chapter, the earliest classification schemes by Wolf (1908), Hubble (1926), de Vaucouleurs (1959), and Sandage (1961, 1975) are purely descriptive. From this came the idea that a galaxy's taxonomy and classification

doi:10.1088/2514-3433/abb602ch2

should be based only on the most obvious features seen in optical light (Sandage 2005). The reason for this philosophy is that no physical understanding of what drives galaxy formation/evolution should be used to develop a classification system. Sandage (2005) puts it this way, "A good classification can drive the physics, but the physics must not be used to drive the classification." It is important to ask if this is the correct way to view this problem, or if indeed the physical mechanisms that form and evolve galaxies should at least be considered in deciding what aspects are used to classify it.

This idea, which follows Hubble's original philosophy in many ways, actually resulted in part from the earliest attempts to classify galaxies through the use of theoretical arguments for galaxy formation developed by Jeans (1919). Hubble avoided theory in some aspects when developing his classification system, however the terms we use for ellipticals, "early-types," and for spirals, "late-types," and the order of the assembly of the Hubble sequence is based on the theoretical ideas of Jeans. It is also important to remember that physical morphology does not necessarily rest upon theory, but empirical observations. Physical morphology in this sense is not a theoretical morphology, but a morphology which traces the most important underlying characteristics of a galaxy. This implies that if we could determine the most important modes seen in structure that drove the formation of galaxies, a classification based on these features would be superior to a subjectively based system.

As described in the first chapter, there are features of galaxies we can observe that make them fundamentally different from other objects that are classified based on their appearance, including stars, plants/animals, and the chemical elements. In many ways galaxies are fundamentally different from these objects as they are not stable or passive systems, nor do they have a well defined life and death as stars or living things do. Another difference is that elements, plants/animals, and many stars spend most of their life in the same form and do not evolve between types. This was in fact a major question when stars were first characterized—it was not clear how the different types evolved into each other and it was quite a dramatic discovery to find that stars did not evolve along the main-sequence, but evolved off of it. The situation for galaxies is different, whereby it has been shown that systems can change their appearance in optical light dramatically in the last 10 Gyr (e.g., van den Bergh et al. 2001; Conselice et al. 2005a; Ravindranath et al. 2006; Mortlock et al. 2013). Essentially, it is the case that galaxies are not static systems and they are always changing and evolving through a complex mixture of stars, gas, dust, dark matter, and from interactions with other galaxies.

It is also not a given that the rest-frame optical light of galaxies, which is what we traditionally use to classify galaxies, reveals the most important features of galaxies. This is yet another reason why the physical properties of galaxies can and should in many cases be used to devise a classification system. There are a couple of good reasons for accepting this. For example, we know that the bulk emission from an average galaxy is in the mid-infrared, and thus if any wavelength should be used to classify galaxies the case can be made that it should be done at the wavelength were most of the light is emitted. It is also the case that the optical light does not display the bulk of the stellar mass of a galaxy, which is better traced in the near-infrared. It can be argued that the most accurate way to classify galaxies is by using a map of the distribution of the total mass, which would be mostly dark matter. A dark matter map such as this, with high enough resolution for classification, does not currently exist, although dark matter profiles have been measured for galaxies. The importance of this is to realize that in blue light, where the Hubble system was and is applied, we are often only examining the brightest stars that often tend to be young.

The other side of this is understanding which properties of galaxies are not important for classification which is just as critical as finding the dominant ones. This can be seen through examining analogies with other areas of science where specific categories of objects are classified. One example of this is in biological taxonomy where features such as hair or eye color, or exact size, are not usual classification criteria. For galaxies we have the situation where features such as rings, bars and even spiral arms are the most notable and obvious features in the structure of a galaxy, but these are not necessarily the most important. Classifications that use these features are purely descriptive without any regard for their being fundamental features. These criteria are also largely arbitrary, as other properties, including the steepness of light profiles, clumpiness of the light, as well as the spokes and spurs seen in spiral arms, etc. were not, and largely are not, used to classify galaxies into various types. The question is: what are the major aspects of galaxies that we should use to characterize them? Part of this is determining what are the major properties which describe galaxies, and if any formation modes can be seen through structure.

If we can determine the major properties that characterize the formation and evolution of galaxies we can then in principle use them to construct a physically motivated classification system. This idea of a physical classification is not a new idea, and many systems were developed to do this, including the early work of Morgan (1958, 1959) who correlated "the forms of certain galaxies and their stellar content as estimated from composite spectra" from data taken within Morgan & Mayall (1957). Further attempts also include van den Bergh's (1960) examination of the correlation of spiral arm structure and the intrinsic brightness of galaxies. Other more modern works classify ellipticals by their structures (Kormendy & Bender 1996), and at using interacting and star formation properties to classify galaxies (e.g., Conselice et al. 2000; Bershady et al. 2000). This latter aspect we will come back to later in this chapter.

2.1 Analyzing the Properties of Galaxies

In this section of this chapter we discuss the best ways in which to determine what are the major galaxy properties. We know that a dominant aspect of galaxies is their stellar mass, which correlates strongly with properties such as star formation, and evolutionary processes are determined mostly by stellar mass (e.g., Grützbauch et al. 2011; Disney et al. 2008). One way this can be shown is to use the correlation of morphology with physical properties such as: Hubble type, size, color, surface brightness, magnitude, stellar mass, internal velocities, HI gas content, and a kinematic index to measure kinematic disturbances. When this is examined through a Principal Component Analysis (PCA) what is found is that Hubble type, color and stellar mass all strongly related (Conselice 2006).

A PCA analysis reduces the number of variables (n) in a set of variables to a set of $m < n$ parameter combinations, thereby simplifying which parameters are needed to accommodate all the variance in galaxies. This process works through the calculation of eigenvectors and eigenvalues for each parameter combination. How this works in practice is that the first eigenvector is a line, which weights different parameters in different ways, such that this line goes through the cloud of n data points, minimizing the variance. From this, the remaining eigenvectors are calculated until all the variance is explained. The numerical values of the eigenvalues reveal how much of the variation is accounted for by each eigenvector within the cloud of points. The eigenvalue must be $\geqslant 1$ for an eigenvector to be considered significant.

Through using a PCA analysis of 22,000 nearby galaxies reveals that most of the variation in properties are accounted for by eigenvectors dominated by: (i) a galaxy's "scale," such as its stellar or total mass, (ii) the spectral type, or star formation rate relative to old stars, and (iii) the degree of dynamical disturbances (Conselice 2006). The eigenvectors and eigenvalues for this analysis taken from Conselice (2006) are given below in Table 2.1.

When a PCA analysis on a galaxy sample using only the parameters of Hubble type (often call a T-type), spectral indices such as $(B-V)_0$ color, a merger index (denoted by I), and stellar mass (M_*) the eigenvectors obtained are displayed in Table 2.1. These eigenvectors in total make up 92.3% of the variance in the properties of galaxies. Due to the expected errors on these measurements, this is likely the limit for meaningful eigenvectors (Conselice 2006). The T-type, stellar mass, and color correlate strongly, and they dominate the first principal component. The I merger index defines the major part of the 2nd principal component, while the 3rd component is dominated by the morphological type, and stellar mass. From this, we can conclude that the morphological type is in general successful as a galaxy classification criterion, and it strongly correlates with color and stellar mass among other similar properties. In Figure 2.1 the correlation between the PCA projections of galaxies onto their three primary eigenvectors is shown. These in general correlate with color or star formation rate versus existing stellar mass (p_1),

Table 2.1. Eigenvectors of Galaxy Properties using PCA with T-type, Color, and M_*.

Property	EV 1	EV 2	EV 3
T-type	−0.57	−0.13	0.60
$(B-V)0$	0.63	−0.04	−0.10
I	0.06	0.96	0.23
M_*	0.52	−0.20	0.76
Eigenvalues	2.0	1.02	0.64
	50.8%	25.4%	16.1%

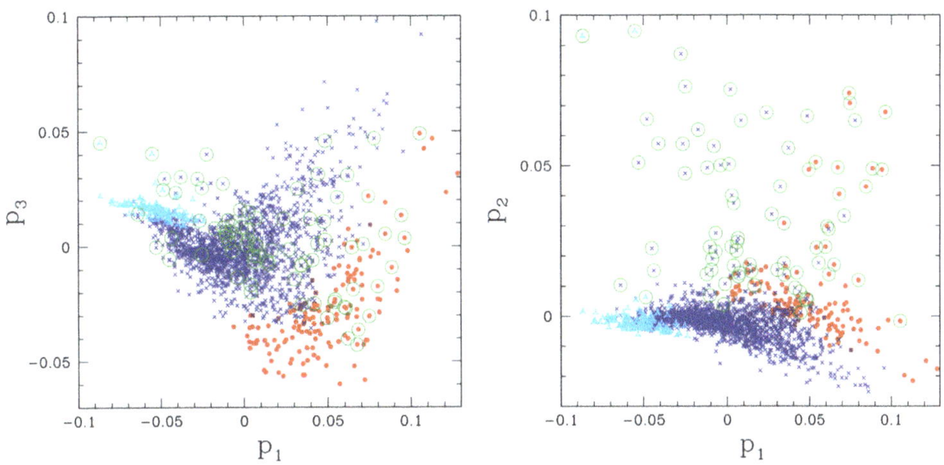

Figure 2.1. Figure showing the three different eigenvectors for the sample of galaxies in Conselice (2006) plotted against each other. As can be seen, the different galaxy types are separated distinctly from one another given by the range of the different colored points. The red solid points are for the ellipticals and S0s, while the blue crosses are discs, and the cyan triangles are irregulars. Green circles are galaxies with an interaction/merger index $I > 2$. The three projected components trace the scale (p_3), star formation (p_1), and interacting/merging properties (p_2) of galaxies.

interactions/mergers (p_2), and stellar mass (p_3). In this figure we label the morphological type of each galaxy, showing that this PCA analysis separates in an impressive way the major galaxy types, including: early-types, disks, irregulars, and merging galaxies. Based on this we can conclude that the fundamental properties for describing nearby galaxies are their scale as represented by mass, recent star formation, and merger activity. This is backed up by observations which show that the color, stellar populations, and star formation rates are highly correlated with the stellar mass of a galaxy. This is in fact one reason why the Hubble sequence itself works well, as the most massive galaxies tend to be elliptical and lack star formation.

From this argument, it is clear that star formation, interactions/mergers, and the masses of galaxies are fundamental parameters that can be used together to form a physical galaxy classification (Figure 2.2). The question becomes whether we can measure these features using just the images of galaxies. In the following subsections we describe these various properties and how they can be measured in a morphological or structural way.

2.1.1 Scale or Mass of Galaxies

The case was made in the previous section that the underlying mass (scale) of a galaxy is a fundamental property, which is responsible for the major component in a PCA analyses of galaxy features (Conselice 2006). While few would argue that the mass of a galaxy is not a fundamental criterion for classification, it is not obvious that it is possible to classify galaxies based on their mass, without direct measurements. Direct measurements can include the stellar mass as measured through the amount of light at various wavelengths being emitted or through dark matter measurements using a variety of techniques include lensing, kinematics or comparison to models (e.g., Conselice et al. 2018).

The question is however if there are any morphological or structural indicators that reflect this primary component in galaxies. If not, then galaxy structure and morphology would be much less a useful tool if the primary aspect for understanding galaxies was not reflected in its structure. There are fortunately structural indicators that exist in optical light that can be used as a proxy for the mass of a galaxy. For example, as discussed in, e.g., Graham et al. (2001) and Conselice (2003) the concentration of a galaxy measured with either parameterized fits to light distributions, or from direct measurements (Bershady et al. 2000), broadly correlate with the stellar and kinematic masses of galaxies. This is true at low redshift, as well as out to $z \sim 1$ (Conselice et al. 2005a, 2005b).

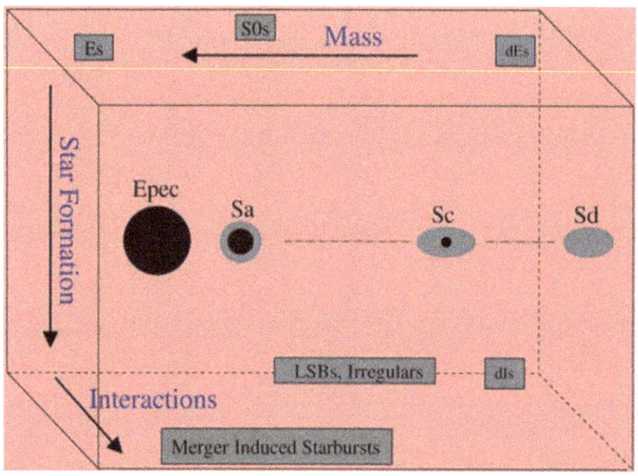

Figure 2.2. The three-dimensional classification volume C^3 which can account for all known galaxy types and perhaps ones that have not yet been discovered. The C^3 parameter space is generalized and idealized in this space and includes the features of mass, recent star formation, and finally the interaction and merger properties of galaxies. Within the C^3 space well known galaxy types are labeled in their appropriate locations in this parameter space. At the top of the classification cube are galaxies without recent star formation. As we go from dEs to S0s to giant ellipticals along the mass sequence we see there is a correlation between mass and morphology. Galaxies with recent interactions/mergers are located further into the third dimension. Along the y-axis, we find systems with a larger amount of recent star formation relative to its stellar mass. This quantity is essentially a measure of the specific star formation rate (sSFR = SFR/M_*). Systems that are low surface brightness disks are located in a region of this space where the star formation is a bit high, and interaction history is low. On the other hand, merger induced starbursts are found in the high star formation, high interaction part of the diagram. Finally, those galaxies classified as irregulars are located toward the bottom right of the cube given that these are typically lower mass star-forming galaxies. Other morphological sequences are projections of this space. This includes the Hubble sequence, which is a one dimension projection of the star formation and interaction axes onto the mass sequence. This classification necessarily replaces idealized Platonic Hubble types with idealized parameters that are not necessarily easy to retrieve, but in principle are measurable.

How much the light in a galaxy appears concentrated toward its center has been used, even if not explicitly stated, since at least Hubble (1926) as one of the main criteria for classifying galaxies. Papers such as Conselice (2003) demonstrate this inherently by showing that galaxy concentration correlates with the bulge to disk ratio of galaxies. The degree of light concentration in a galaxy is almost ubiquitous in galaxy classification such as in papers such as de Vaucouleurs (1959) and Morgan (1959). This quantity continues as a major aspect for galaxy classification using quantitative structural measurements, including fitting bulge to disk ratios which is also a proxy for galaxy concentration. More recently, Driver et al. (2006) show that the light concentration and color are fundamental properties separating galaxies into early and late types. Along similar lines, Blanton et al. (2005) find that color and luminosity are the dominant properties of galaxies and are the ones most useful for separating galaxies into various types.

Although bright Hubble types correlate with the mass of galaxies, they do not account for all galaxy properties needed for a full classification. In fact, it is clear that there is a wide range of properties at each Hubble type (Roberts & Haynes 1994; Kennicutt 1998). This is a limitation of the Hubble sequence, which simplifies classifying galaxies into a single 1-dimensional classification that hides diversity in other properties not related to the overall scale. One example of this is the fact that late-type galaxies, which generally have low masses, and lower bulge to disk ratios, and low light concentrations, are blue is a result of evolutionary processes at work within these systems. The colors of these galaxies are not due to an inherent success of the classification scheme. It is somewhat of a coincidence that Hubble types correlate well with star formation properties in the nearby universe, because this correlation is not seen at higher redshifts where classified ellipticals are often blue in color and star-forming (e.g., Stanford et al. 2004; Bundy et al. 2005; Conselice et al. 2011). The agreement between star formation history and Hubble type breaks down completely at $z > 1$, where the Hubble sequence itself is rarely seen (see Chapter 4).

2.1.2 Star Formation

As discussed in Chapter 1 and Section 2.1 there is a morphological dichotomy between galaxies that are dominated by recent star formation and those that are not. The 2nd principal component discussed above reveals that the color of a galaxy is the major contribution to one of the most significant principal components. This is not accounted for by the Hubble type or overall morphology as classification as an elliptical or a spiral does not necessarily reveal this property. That is, there is a range of physical properties within each Hubble type that is not related directly to the star formation rate as opposed to the amount of mass that is already present within the galaxy. In other words, the Hubble sequence is not a complete description, as within each type there is a large range of star formation histories.

We can see this in terms of data as galaxies at a given morphological type have a range of colors, or star formation histories. Therefore, if we knew the current star formation rate of a galaxy it would be a very powerful classification parameter. The current star formation rate of a galaxy can be derived from a flux, such as ultraviolet, Hα, or far-infrared fluxes (Kennicutt 1998), although these measurements can be hard to obtain for large samples of galaxies. In terms of morphology and structure we want to know if there is a feature that can be used to estimate the star formation properties of galaxies based solely on its optical imaging.

The answer to this is that it is possible to use structure as a measure of the ongoing star formation rate. One example of this is the grand design versus flocculent spirals that have different star formation rates (e.g., Elmegreen & Elmegreen 1987). This is another success of the method for using galaxy structure as a way to probe the properties of galaxies. Furthermore, we can quantify this in more detail as a measure of star formation rate using high-frequency structural features produced by young stars (e.g., Isserstedt & Schindler 1986; Conselice 2003). The clumpiness (or smoothness) index, *S*, (Conselice 2003) can be a proxy for measuring the unobscured current star formation rate in galaxies (Figure 2.3; see next chapter).

2.1.3 Galaxy Mergers/Interactions

Galaxy mergers and interactions have already been discussed in this book, and they are indeed a running theme throughout the entire book as they touch on so many issues. Clearly, this is an important process in galaxy formation, and is one of the main things that we can certainly visually see when examining galaxies. Two galaxies merging, such as the example in Figure 1.6, are quite visually striking and obviously much different from ellipticals and spirals that constitute the traditional view of galaxies. The question is—how important are mergers in the galaxy formation process? This is an issue that we will address in detail later in this book. The question we address now is whether interactions are one of the major features needed to characterize galaxies.

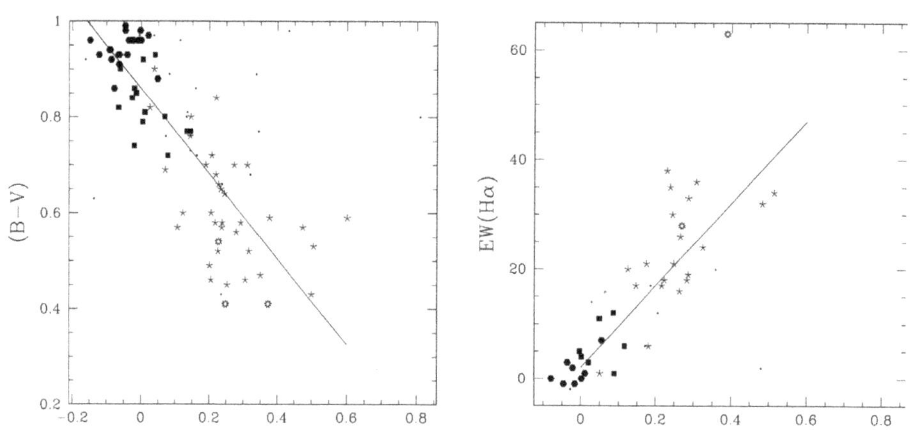

Figure 2.3. The relation between star formation rate parameters: color and equivalent width of the Hα line correlating with the clumpiness index as measured from the images of galaxies. This index is introduced and explained in more detail in the next chapter. Reproduced from Conselice (2003). © 2003. The American Astronomical Society. All rights reserved.

Similar to the scale of a galaxy and its star formation rate, galaxy mergers and interactions are in fact the third major important aspect for describing galaxy properties. This may come as some surprise as mergers are rarely found in the local galaxy population, but both simulations, theory of structure formation, and observations (e.g., Conselice et al. 2003) show that galaxy mergers are a major aspect for the early formation of galaxies. For example, ellipticals and massive bulges are likely the results of major mergers between galaxies that have similar masses. However, past merger events for any one galaxy are very difficult to know about with certainty, especially ones that would have gone through violent relaxation, by which the original orbital information of the stars before the merger is lost. On the other hand, more recent interactions between galaxies will always leave obvious morphological and structural signatures that will change the way galaxies appear (see Chapter 3). These mergers will also clearly have an effect on their internal evolution, as has been shown in various studies.

2.2 A New Classification System: The Classification Cube

We show in Figure 2.2 how different types of galaxies can be classified into a three-dimensional parameter space defined by galaxy mass, recent star formation, and interactions. As shown, the x-axis signifies the mass, while the y-axis represents the recent star formation, and the z-axis shows the degree of interactions and mergers with other galaxies. In many ways this can be seen as a system with a main axis, which is mass, and two perpendicular axes that correlate with the rate for how that main axis changes—one due to star formation and one due to the amount of mass brought in through mergers or removed through interactions (which is much less common). The main axis is then the integral of the other two axes through time.

The system presented in Figure 2.2 simply account for all the major galaxy types known in the local and distant universe. For example, ellipticals, with no sign of recent interactions or star formation, occupy the part of the parameter space where high-mass, low star formation and low interaction galaxies are found. The very top of this volume contains the S0s as well as the dwarf ellipticals/spheroidals. This is because we know that S0s have little to no recent star formation, but are on the other hand usually lower mass systems than ellipticals. As such, the S0s are found in the same star formation and interaction part of this space as the ellipticals but lower down the mass axis. The dwarf ellipticals contain the most extreme low masses, and as such they are found in the far end of the mass sequence. On the other hand, the dwarf irregulars have a similar bulk internal structure and mass to the dEs, but they are undergoing intense star formation, and thus we would place this at the same mass and interaction space as dEs, but separated in star formation space (Figure 2.4).

The classification cube (abbreviated C^3) is not a totally new way of looking at galaxies, as various aspects of it have been discussed in different forms. One example of this is the work of van den Bergh (1976) who proposed a new revision of the Hubble scheme such that S0s form a parallel sequence to the spirals, instead of being located in between ellipticals and spirals as is the case within the original Hubble

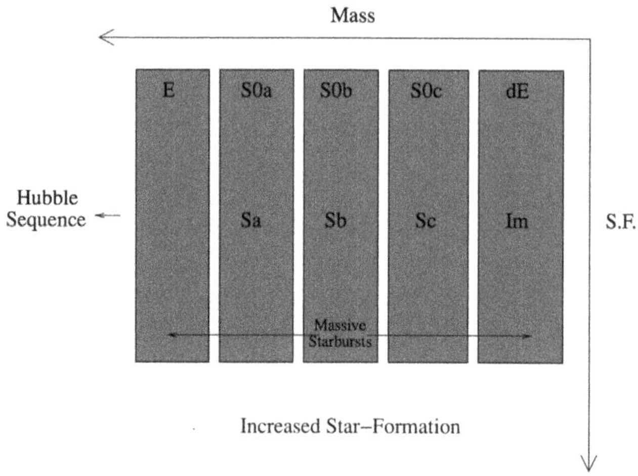

Figure 2.4. The system in Figure 2.2, but now projected onto star formation/mass plane. As shown, the Hubble sequence is labeled, and includes galaxies with a range of masses, modest star formation with no distinction made for interactions. Those galaxies that are traditionally classified as ellipticals, S0s, dEs, and dwarf spheroidals are located at the bottom of the star formation sequence. Furthermore, unusual galaxies, such as massive starbursting systems are those systems located at the high star formation area of this projection.

sequence. This classification scheme also contains the so-called "anaemic" galaxies which contain little gas or star formation. Galaxies in this system are classified in terms of their bulge to disk ratios, making it very similar in some ways to the star formation/mass plane within C^3.

Other galaxies also fit into the C^3 system. This includes the low surface brightness galaxies, which have little evidence for recent interactions with other galaxies (Figure 2.5). They also tend to have lowish star formation histories that are still actively forming stars. In this sense, LSB galaxies are found in the same mass sequence as the spirals, but will be found in the low interaction areas of the interaction sequence. Boxy versus disky galaxies can also be separated within this C^3 volume, as each one has a different past history of galaxy mergers, but in some cases these were some time ago.

Galaxies undergoing intense star formation, or the so-called starbursts, which are usually systems induced by major mergers, are located at the high end of the interaction and star formation sequence. Those mergers which are major and undergoing violent relaxation, such as Arp 220, containing a proto-spheroidal component and thus a steep light profile (Wright et al. 1990), would be classified toward the high-mass end of the mass sequence. This system is based on the results from an analysis of the properties of galaxies and their principal components showing that these are the three major properties of galaxies that describe most of the variance in all of their major measurable features. If one could measure the star formation, merging, and masses of a galaxy it would be possible to classify a system without recourse to structure or morphology, but often this is not possible. For example, we could carry out these classifications directly by measuring star

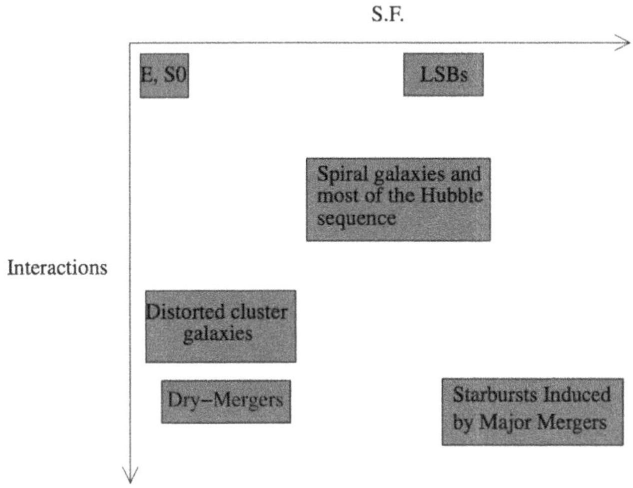

Figure 2.5. Similar to Figure 2.4 but a projection of the star formation/interaction plane of Figure 2.2. In this projection, we can see different galaxy types that are not represented in the star formation, mass plane. As can be seen, there are several classes of galaxies separate along this plane perpendicular to the mass sequence. Galaxies which would be described as low surface brightness (LSBs) galaxies are those with modest star formation and very little recent interaction history. As displayed, the ellipticals and S0s have little recent star formation or interactions, and no recent interactions/mergers. Several other galaxy types, such as distorted cluster galaxies undergoing little star formation due to gas removal by the cluster, dry-mergers, and starbursts induced by major mergers, are also labeled.

formation rates based on broad band fluxes and emission lines, masses based on dynamics or spectral energy distributions, and mergers/interactions based on an asymmetry index or a dynamical indicator (see next chapter).

2.2.1 Idealized Classifications

When early classification systems such as Hubble's were invented, they were made to account for specific types of galaxies that looked similar to each other. We now know however that these Hubble types are largely idealized classes of galaxies and within each class there is a broad range of real galaxies with various physical properties. Few galaxies fit the ideal Hubble types, although massive ellipticals come the closest as they are generally all similar to each other, nor is any one galaxy exactly the same as any other. This is largely the result of the chaotic or stochastic interaction and star formation history galaxies undergo which cannot easily be retrieved or inferred.

The C^3 idea, which we use in one way or another through most of this book as the philosophical underpinning of much of what we explain, replaces idealized galaxies with parameters that are in some ways idealized measurements. However, they are ideal because of technical limitations, not because these features do not exist. It is not always easy to determine the star formation history of a galaxy, nor is it always accurate, with often large uncertainties. As we discussed, the method for classifying the star formation properties of a galaxy does not have to be based on morphology, but can also be done through direct measurements, but in these cases, and in the morphological case, currently we can only know with some uncertain value where a particular galaxy falls in the idealized parameter space. This however could easily change once we have better measurements of these features in the future and can measure with precision the actual star formation rate.

This is also true when measuring the masses and merger histories of galaxies. Many galaxies have some clues leftover from merger events, but these can be quite diverse and difficult to measure or see. On the other hand, structural methods, such as the asymmetry parameter (Chapter 3; Conselice et al. 2000), can be used to determine the presence of ongoing major mergers. The history of past interactions is more difficult to retrieve or to measure accurately. There are ways of doing this however. One of these is to examine the outer parts of galaxies where the dynamical times are long, up to 10 Gyr, and where many apparently normal galaxies show evidence for merger activity in the form of tidal features (Schweizer & Seitzer 1988). We can find other signs of past mergers including distorted HI halos surrounding galaxies, galaxy core depletion from merging black holes (Graham 2004), and distorted rotation curves (Swaters et al. 1999) or line profiles (Haynes et al. 1998). These are all potentially remnants of accretion, merging or interactions with other galaxies, and can be used to quantify the interaction history as opposed to the primary structure, which gives us the recent merger and interaction history.

2.2.2 A Theoretical View of Galaxy Classification

Above we showed that three parameters of galaxies: mass, star formation, and interactions are the most important features of galaxies, and in terms of parameter space can be used to classify and trace the evolution of galaxies. As we have also shown, these parameters have a direct correlation with the semi-irreversible processes of mass assembly or removal, and the star formation rate. As discussed above, these features can often be measured through stellar light distributions. In this section, we give a continuity equation derivation for how this can be done in terms of structure (see also appendix B of Conselice 2003).

We start by writing the basic initial baryonic mass of a galaxy after its formation, ($\mathrm{M}_{0,\mathrm{B}}$) as,

$$\mathrm{M}_{0,\mathrm{B}} = \mathrm{M}_{0,\mathrm{S}} + \mathrm{M}_{0,\mathrm{G}}, \tag{2.1}$$

whereby the values $\mathrm{M}_{0,\mathrm{S}}$ and $\mathrm{M}_{0,\mathrm{G}}$ are the initial stellar and gas masses. Mass will accrete or be removed from a galaxy due to interactions within a cluster, for example, as time goes on. After a time t, the mass added to or removed from the system by interactions and mergers, denoted by $M'(t)$ is given by,

$$M'(T) = M'_{\mathrm{S}}(T) + M'_{\mathrm{G}}(T) = \int_0^T \dot{M}(t)\mathrm{d}t = \int_0^T \dot{M}_{\mathrm{S}}(t)\mathrm{d}t + \int_0^T \dot{M}_{\mathrm{G}}(t)\mathrm{d}t, \tag{2.2}$$

with the value $\dot{M}$ the total mass accretion, or removal, rate as a function of time, and T is the age of the galaxy when observed. Further dividing the mass into a stellar $\dot{M}_{\mathrm{S}}$ and gas $\dot{M}_{\mathrm{G}}$ components, we can determine how these quantities change with time. Obviously, some gas will be converted into stars at each time, t. We consider the rate of this occurring using a conversion rate, $\alpha(t)$, such that the gas mass converted into stars, $\tilde{M}_{\rm S}$, is given by

$$\tilde{M}_{\rm S}(T) = \int_0^T \alpha(t) M_{\rm G}(t) {\rm d}t, \tag{2.3}$$

where $\alpha(t)$ and $M_{\rm G}$ are the gas conversion rate, and gas mass as a function of time. We can then write the total stellar and gas masses as a function of time by:

$$M_{\rm S}(T) = {\rm M}_{0,{\rm S}} + \int_0^T (\dot{M}_{\rm S}(t) + \alpha(t) M_{\rm G}(t)) {\rm d}t, \tag{2.4}$$

for how the stellar mass changes with time (t) up to some final time (T). For the gas mass we have a similar equation:

$$M_{\rm G}(T) = {\rm M}_{0,{\rm G}} + \int_0^T (\dot{M}_{\rm G}(t) - \alpha(t) M_{\rm G}(t)) {\rm d}t. \tag{2.5}$$

The three C^3 axes can be represented by three terms in these equations. The interaction measurement based on stars is represented by $M_{\rm S}'(t)$, and the star formation parameter by $\tilde{M}_{\rm S}$. To simplify this we can create dimensionless parameters that in principle will correlate with stellar light features. The integrated interaction, $\overline{I(t)}$ and star formation $\overline{SF(t)}$ strengths are dimensionless quantities that are meant to correlate with observable features in structure, which are also non-dimensional quantities. These terms are weighted by the final mass, and time and can be written as:

$$\overline{I(T)} = \int_0^T \frac{t}{T} \frac{|\dot{M}_{\rm S}(t)|}{M_{\rm S}(T)} {\rm d}t \tag{2.6}$$

$$\overline{SF(T)} = \int_0^T \frac{t}{T} \frac{\alpha(t) M_{\rm G}(t)}{M_{\rm S}(T)} {\rm d}t. \tag{2.7}$$

These morphological parameters are quantities that correlate with the appearance of a galaxy. We weight these by time since star formation and interactions in the more distant past will have less strong of an effect on the appearance of a galaxy. Simulations would have to be applied to determine if this is indeed a linear trend or if there should be an exponential term on the time here, which could differ for the star formation and merging parameters. More massive galaxies will also be less affected by lower amounts of gas converted into stars, and stars lost or gained from galaxy interactions. Therefore these quantities are also weighted by the final mass of the galaxy, which we use the stellar mass as a proxy for, although the total mass could be the normalizing factor.

By following these equations and the $\overline{I(t)}$ and $\overline{SF(t)}$ terms, the main galaxy formation processes of star formation and interactions/mergers can be measured, or at least inferred through optical light. In fact, these terms correlate with the asymmetry and clumpiness indices that we describe in more detail in the next chapter. When a galaxy undergoes recent star formation it will contain clumpy light, occupied by clusters of young stars, which then dissipate and blend into the general light of a galaxy after some time. In a similar way, a galaxy undergoing a merger or interaction will appear distorted and lopsided for some time with the strength of that feature in morphology dependent on both the mass of the galaxy and how long it has been since the event occurred. In summary, by measuring or estimating the degree of: disturbances, clumpy structures, and how concentrated a galaxy is (i.e., bulge to disk ratios) we can classify galaxies into the C^3 volume using optical light. This approach has been successful for classifying both nearby (e.g., Conselice 2003) and distant galaxy populations (Conselice et al. 2004; Grogin et al. 2005; Papovich et al. 2005; Mortlock et al. 2013).

2.2.3 Could There Be Unidentified Galaxy Types?

The C^3 system is quite effective at classifying all known galaxies, but there are also classification "corners" where no galaxies are known to exist currently. These in principle could be new types of galaxies that have not yet been discovered to date. One of these types are galaxies that are massive, have a low interaction history and a high star formation rate. Many galaxies have these properties at high redshift which are likely ellipticals or disks in formation (e.g., Margalef-Bentabol et al. 2016). These systems could originate from gas accreted onto elliptical-like systems to form disks, or large disk galaxies in formation. It is possible that these systems did exist but have been destroyed by later mergers. It is also likely that these systems do exist but at earlier phases of the universe than nearby galaxies. In many ways, these would be galaxies that would appear if "monolithic collapse" were occurring to produce galaxies, rather than the hierarchical assembly that we know most massive systems form through.

Another interesting type not yet identified are galaxies at various mass ranges, but which have a low star formation rate, but a high interaction strength. A good candidate for these galaxies are the population of distorted, compact dwarfs found in clusters (Conselice & Gallagher 1999) with little star formation. These are examples of galaxies whereby the gas has been removed by the intracluster medium and interactions have destroyed structures, perhaps by fast galaxy harassment processes. Another good example of this are the so-called "dry" mergers between spheroids which contain little amounts of gas (e.g., Hernández-Toledo et al. 2006). In this situation, we have a high interaction/merger history, but a low star formation one. There are even more possible galaxy populations that can be hypothesized by considering missing areas of the C^3 cube shown in Figure 2.2.

2.3 Summary

This chapter forms the basis for much of the understanding of how galaxy structure can reveal important information about galaxies, both nearby and in the distant universe. It can be shown through various arguments that the major properties of galaxies are their mass, their ongoing star formation rate, and their merger/interaction history. In this chapter we demonstrated that these three parameters or features can be used to classify all known galaxy types. Therefore galaxy classification is a 3-parameter cube, called C^3 which has "idealized" axes made of these parameters. In the following chapters we explain how to measure these features through galaxy structure, and how these features are revealing the formation history of galaxies.

References

Bershady, M., Jangren, A., & Conselice, C. J. 2000, AJ, 119, 2645
Blanton, M. R., Eisenstein, D., Hogg, D. W., Schlegel, D. J., & Brinkmann, J. 2005, ApJ, 629, 143
Bundy, K., Ellis, R. S., & Conselice, C. J. 2005, ApJ, 625, 621
Conselice, C. J., & Gallagher, J. S. 1999, AJ, 117, 75
Conselice, C. J., Bershady, M. A., & Jangren, A. 2000, ApJ, 529, 886
Conselice, C. J. 2003, ApJS, 147, 1
Conselice, C. J., Bershady, M. A., Dickinson, M., et al. 2003, AJ, 126, 1183
Conselice, C. J., Grogin, N. A., Jogee, S., et al. 2004, ApJL, 600, L139
Conselice, C. J., Blackburne, J. A., & Papovich, C. 2005a, ApJ, 620, 564
Conselice, C. J., Bundy, K., Ellis, R. S., et al. 2005b, ApJ, 628, 160
Conselice, C. J. 2006, MNRAS, 373, 1389
Conselice, C. J., Bluck, A. F. L., Ravindranath, S., et al. 2011, MNRAS, 417, 2770
Conselice, C. J., Twite, J. W., Palamara, D. P., et al. 2018, ApJ, 863, 42
de Vaucouleurs, G. 1959, Handb. der Physik, 53, 275
Disney, M. J., Romano, J. D., Garcia-Appadoo, D. A., et al. 2008, Natur, 455, 1082
Driver, S. P., Allen, P. D., Graham, A. W., et al. 2006, MNRAS, 368, 414
Elmegreen, D. M., & Elmegreen, B. G. 1987, ApJ, 314, 3
Graham, A. W., Trujillo, I., & Caon, N. 2001, AJ, 122, 1707
Graham, A. W. 2004, ApJL, 613, L33
Grogin, N. A., Conselice, C. J., Chatzichristou, E., et al. 2005, ApJL, 627, L97
Grützbauch, R., Conselice, C. J., Bauer, A. E., et al. 2011, MNRAS, 418, 938
Haynes, M. P., Hogg, D. E., Maddalena, R. J., Roberts, M. S., & van Zee, L. 1998, AJ, 115, 62
Hernández-Toledo, H. M., Avila-Reese, V., Salazar-Contreras, J. R., et al. 2006, AJ, 132, 71
Hubble, E. 1926, ApJ, 64, 321
Isserstedt, J., & Schindler, R. 1986, A&A, 167, 11
Jeans, J. H. 1919, Problems of Cosmogony and Stellar Dynamics (Cambridge: Cambridge Univ. Press)
Kennicutt, R. C. 1998, ARA&A, 36, 189
Kormendy, J., & Bender, R. 1996, ApJ, 464, 119L
Margalef-Bentabol, B., Conselice, C. J., Mortlock, A., et al. 2016, MNRAS, 461, 2728
Morgan, W. W., & Mayall, N. U. 1957, PASP, 69, 291
Morgan, W. W. 1958, PASP, 70, 364
Morgan, W. W. 1959, PASP, 71, 394
Mortlock, A., Conselice, C. J., Hartley, W. G., et al. 2013, MNRAS, 433, 1185
Papovich, C., Dickinson, M., Giavalisco, M., et al. 2005, ApJ, 631, 101
Ravindranath, S., et al. 2006, arXiv: astro-ph/0606696
Roberts, M. S., & Haynes, M. P. 1994, ARA&A, 32, 115
Sandage, A. 1961, The Hubble Atlas of Galaxies (Washington, DC: Carnegie Institute of Washington)
Sandage, A. 1975, in Galaxies and the Universe, ed. A. Sandage, M. Sandage, & J. Kristien (Chicago, IL: Univ. Chicago Press)
Sandage, A. 2005, ARA&A, 43, 581
Schweizer, F., & Seitzer, P. 1988, ApJ, 328, 88
Stanford, S. A., Dickinson, M., Postman, M., et al. 2004, AJ, 127, 131

Swaters, R. A., Schoenmakers, R. H. M., Sancisi, R., et al. 1999, MNRAS, 304, 330
van den Bergh, S. 1960, ApJ, 131, 215
van den Bergh, S. 1976, ApJ, 206, 883
van den Bergh, S., Cohen, J. G., & Crabbe, C. 2001, AJ, 122, 611
Wolf, M. 1908, PAIKH, 3, 109
Wright, G. S., James, P. A., Joseph, R. D., et al. 1990, Natur, 344, 417

Christopher J Conselice

Chapter 3

Measuring Galaxy Structures

This chapter describes modern attempts to measure the structure of galaxies by which is meant the sizes, quantitative structures and morphologies of galaxies. This chapter furthermore will review the current state of the art for measuring these features within galaxies in the local and distant universe using the best and most up to date results. It also discusses how galaxy structure relates to their detectability, sizes, and surface brightness. There are in fact many ways to measure galaxy structures, and this chapter is meant to be a detailed description of the methods and techniques to do this, and the limitations of using them. As this book is focused on the evolution of galaxy structure with time, this chapter is an important one for what comes later where many of these tools are used to derive the structural and then galaxy evolution properties up to $z \sim 6$.

As discussed in detail in Chapter 1, the apparent morphology is one way in which to characterize the structures of galaxies, and this is by far the most common and historical. This is still being done today by many, and has even had a resurgence of popularity with projects such as Galaxy Zoo, where so-called Citizen Scientists can log in and classify galaxies. These classifications by tens of thousands of volunteers are then collated to produce a catalog of galaxy morphologies (Lintott et al. 2011). Already this approach has led to the classification of over a million galaxies with assistance from over 100,000 volunteers. These original Galaxy Zoo classifications were based on imaging done by the Sloan Digital Sky Survey (SDSS). This is still in many ways the largest and most homogeneous imaging data set of galaxies to date.

However, using visual human-based morphologies to determine the structure of a galaxy is an inherently limited process. While these type of measurements can tell us how a galaxy fits into a predetermined system of taxonomy, it is not clear if the characteristics common for these morphologies are the ones that are most correlated with the actual physical properties of galaxies (see Chapter 2; Conselice 2006a). The analogy here would be using properties of living creatures to classify them such as height and weight or coloring of fur or skin which are not fundamental properties of

doi:10.1088/2514-3433/abb602ch3

a species. It is unlikely that the dominant methods we use to classify galaxies within the so-called Hubble system is strongly biased in this way, but it can be improved to better represent the most important features of galaxies (see Chapter 2). Even after a century this issue is still debated among those who study galaxy structure.

The most basic way of categorizing galaxies in a quantitatively way is by measuring the flux of a galaxy accurately. This was done starting with work with photomultipliers and photometers by Albert Stebbins at the University of Wisconsin in the 1930s. This early quantitative photometry was mostly done for stars, but was expanded to include galaxies not long after its introduction. Electronic devices have always been used to obtain accurate measures of photometry given the inherently non-linear and difficult to calibrate nature of photographic plates.

The apparent flux of a galaxy f is determined by the intrinsic luminosity of that galaxy (L) and its distance from us (D) through the formula:

$$f = \frac{L}{4\pi D^2}. \tag{3.1}$$

Photometers are not used much in astronomy anymore and this technology is not familiar to many astronomers, but they can be thought of as a one-dimensional imaging device or electronic camera, whereby the entire light of a star or galaxy is measured in one go. From this early photometric work it was clear that galaxies had a range of fluxes and magnitudes, and thereby also a range of surface brightness. This was, in a real way, the first success in quantifying the properties of galaxies. However, it is important to discuss and realize that galaxy light at different wavelengths probe different aspects of the underlying stars within these galaxies. Galaxies observed in bluer light will have images dominated by younger, brighter stars, whereby redder light shows the location for more evolved stars (this issue is described more in Chapter 8). Typically, the redder light shows the underlying stellar mass much better than blue light, as redder light probes lower mass stars which due to the shape of the initial mass function, dominates the total mass of stars within galaxies.

It was not until the photometric work of de Vaucouleurs (1948) that the structures of galaxies were first quantified. In this work de Vaucouleurs measured the flux arising from massive elliptical galaxies at various radii from their centers. He showed that among these bright elliptical galaxies there was an apparent universal trend such that the light decreased as an exponential of the ratio of the distance divide by a scale factor to the one fourth power. This is now known as the de Vaucouleur's law and it is still used as a good representation of the light profile for massive elliptical galaxies. As opposed to the de Vaucouleur's law (sometimes called the $r^{1/4}$ law), disk galaxies have a light profile which is exponential. This was the beginnings of quantifying galaxy structures and it is worth describing this in detail as it is still a major way in which galaxy structure is quantified and studied. We later describe non-parametric ways to quantify the structures of galaxies that do not depend upon an assumed underlying light structural profile.

3.1 Parametric Measurements of Structure

As described above, the first attempt to quantify galaxy structures was through the measurement of integrated light profiles. These light profiles are measured by taking the average intensity of the light from a galaxy at a given radius. From this, a measurement and then fitting of this intensity as a function of radius can be measured. Because the light profile is described by a predetermined fit, these are known as parametric structural measurements, as opposed to non-parametric measurements which are described later in this chapter. The formula for this was also first described in detail by de Vaucouleurs (1948) who measured these properties for ellipticals and proposed a fitting form, given in a modern form by:

$$I(R) = I_e \times \exp(-7.67 \times R/R_e^{1/4} - 1), \tag{3.2}$$

where the value of R_e is the effective or half-light radius. This form for the surface brightness has been shown to be a very good way to fit moderately bright to very bright and massive ellipticals. It however does not succeed in fitting well the very massive systems, nor does it fit very well lower mass systems that might be classified as ellipticals. It is in many ways an empirical fit to the observations rather than one based on any fundamental theory. There are also issues with surface brightness limits and the use of this type of fitting. If we were to use standard fitting routines we would find different answers at times depending on how far out in the galaxy this profile, or similar ones, were measured. This is something to be aware of and does not in itself take away from the usefulness of parametric fitting for describing the structure of a galaxy.

In some ways, a more universal method to fit light profiles of galaxies are to carry out this fitting with a more generalized form of the de Vaucouleur's law, which is often called the Sérsic profile (Sérsic 1963). The so-called Sérsic index n is the major way in which galaxy light profiles are characterized in a parametric way. The general form for this is similar to the de Vaucouleur's profile and is given by:

$$I(R) = I_0 \times \exp(-b(n) \times R/R_e^{1/n} - 1), \tag{3.3}$$

such that the profile shape is described by the Sérsic index, n, and the value of $b(n)$ is determined such that R_e is the effective radius, which by definition contains half of the light within the galaxy. Another special case of the Sérsic index is for exponential disks whereby $n = 1$, giving a simple exponential form for the decline in the intensity as a function of radius. The fitted values of n and R_e are often used to characterize galaxies, both nearby and in the distant universe, and in many ways are among the most standard and commonly used structural parameters for galaxies. However, as mentioned already there can be issues with measuring them within situations of various signal to noise ratios and resolutions (e.g., Graham et al. 2005). The Sérsic profile has however been used to characterized many galaxies in many environments (e.g., Kormendy et al. 2009), including attempts to describe more distant galaxies (e.g., Buitrago et al. 2013). Descriptions for how the Sérsic profile can be used in ways beyond structure and evolution, see papers including, e.g., Kormendy & Djorgovski (1989), Allen et al. (2006), and Simard et al. (2011).

Galaxy two-dimensional profile fitting is now becoming very popular, as the 1-D fitting mentioned above cannot account for features such as rings, bars, and other asymmetries. There are also two component galaxies that should be fit and can be fit better with a "bulge+disk" model whereby the bulge is the central component and the disk is the outer component. The ratio of these two components is often denoted as the bulge-to-disk ratio (B/D), or the bulge-to-total ratio (B/T). These ratios are designed such that at a value = 1 the galaxy is completely elliptical or bulge dominated and for values = 0, the galaxy is dominated by the disk, with no elliptical or bulge component. There are various interpretations of these values, such that an elliptical or massive bulge form in a merger and the disk was later accreted. This is the fundamental idea behind galaxies within a morphological view of galaxy formation that still persists until today (e.g., Steinmetz & Navarro 2002).

These various forms of the Sérsic, exponential and de Vaucouleurs profile are fit through codes such as IMFIT by Erwin (2015), GALFIT by Peng et al. (2002), as well as GIM2D by Simard et al. (2011) among others. These codes allow for a simple and quick method to measure the light profiles and radii of many galaxies, providing data for understanding the evolution of galaxy structure. This allows for the measurements of light components at high-z, although these codes have limitations such as a constant ellipticity assumption within a given component, however they are sufficient for gross measures of galaxy structure.

3.2 Size Measurements

The sizes of galaxies are perhaps the most basic quantitative structural feature although measurements of galaxy sizes is a tricky issue. The main problem is that galaxies have no well-defined edges, and in all cases their light dissipates into the background light of an image. Therefore, measurements such as the effective, or half-light radius, which is used in the Sérsic and de Vaucouleurs profile fitting is biased by the fact that not all the light is ever obtained within an observed galaxy profile. Nonetheless, it is these radii which are used most prominently to trace the evolution of sizes within galaxies (e.g., Trujillo et al. 2007; Buitrago et al. 2008).

This radius is also used very often to determine the sizes of nearby galaxies from which scaling relationships such as the fundamental plane and the size-mass relation are derived and characterized (e.g., Lange et al. 2015). This is also true for early measurements of galaxy radii, which were based on properties of photographic plates (e.g., Redman 1936). One important issue with using a radii fit from a function is that the surface brightness dimming, which scales as $\sim(1 + z)^4$, will make it progressively more difficult to see the outer portions of a galaxy. This means that when the same galaxy is made to appear further and further away, its outer portions become effectively invisible, and the resulting fit to the surface brightness profile can be altered and incorrect, giving radii values that are systematically off (e.g., Graham et al. 2005; Buitrago et al. 2008).

An alternative way of measuring the radius of a galaxy is to use the so-called Petrosian radius, which is based on using the ratio of two surface brightness within a galaxy, and is to first-order independent of distance (e.g., Bershady et al. 2000; Whitney et al. 2019). The reason for this is that because it is defined as a ratio, any effects due to cosmological dimming will be produced on both measures of surface brightness thereby preserving this ratio. The Petrosian or SB ratio method for measuring the radii of galaxies is much less influenced by surface brightness dimming than other methods. It is therefore very useful for measuring the same physical parts of galaxies at different distances (e.g., Petrosian 1976; Bershady et al. 2000; Graham et al. 2005; Whitney et al. 2019).

The Petrosian radius is given by the location in the galaxy where the η values reaches some level. The mathematical form for this value is given by:

$$\eta(R) = \frac{I(R)}{\langle I(<R)\rangle}, \tag{3.4}$$

where $I(R)$ is the surface brightness at a radius R, and $\langle I(<R)\rangle$ is the surface brightness measured within that radius R. As we discuss later in this chapter, non-parametric morphologies and structural parameters such as the CAS parameters are measured at a radius which corresponds to the location usually where $\eta(R) = 0.2$, or a small multiplicative factor of this radius (often 1.5 times) is used (e.g., Bershady et al. 2000; Conselice 2003; Lotz et al. 2004).

3.3 CAS Parameters—Non-Parametric Morphologies

This section introduces and goes into depth about the use of non-parametric structural indices. These allow us to characterize a galaxy's structure without any underlying assumption about its form, which is done when fitting parametrically. These morphological parameters are often very simple to use and apply to galaxies, and they provide a slew of numerical ways to categorize various aspects of the structures of galaxies. The major way in which these non-parametric indices are measured is through the CAS (Concentration, Asymmetry, Clumpiness) system. Figures 3.1 and 3.2 shows graphical examples for how the CAS parameters are measured and correlate with physical parameters such as color.

3.3.1 Asymmetry Parameter

The use of quantitative non-parametric indices for studying galaxy structures was first carried out by Schade et al. (1995). This study introduced the asymmetry index as a method for quantifying how asymmetric a galaxy is. Later this was applied to distant galaxies by Abraham et al. (1996). Both of these studies used Hubble Space Telescope (HST) imaging after the 1993 refurbishment mission. It was not well understood in these early days how this index worked for finding distant mergers, and whether it was finding unique systems or a phase of galaxy evolution.

The first attempt to calibrate the asymmetry index was performed in Conselice (1997) using 43 nearby galaxies to determine how the asymmetry correlates with other galaxy properties such as color and star formation rate. The asymmetry index was later defined with various improvements whereby today it is one of the most

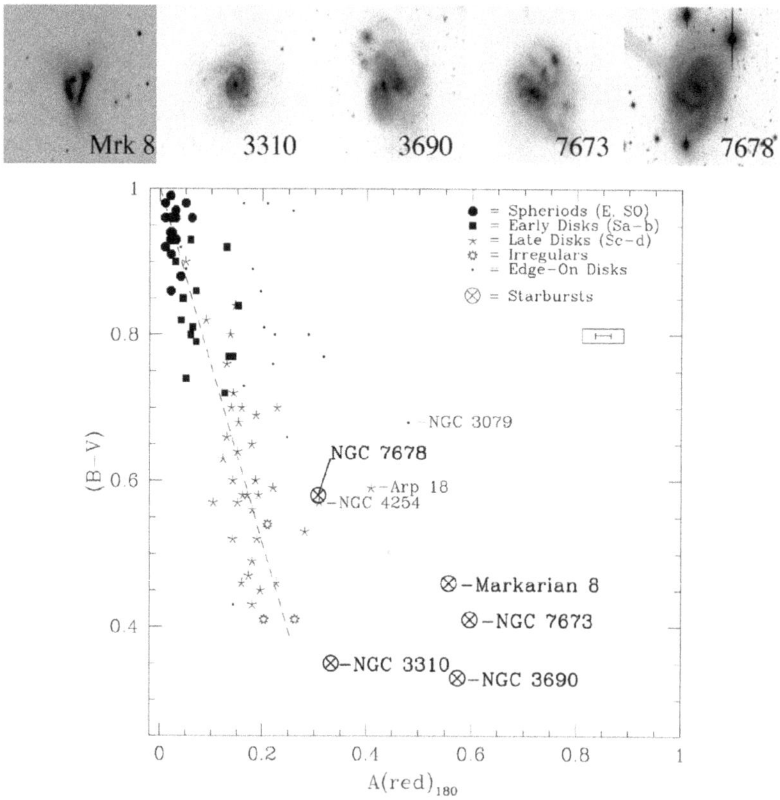

Figure 3.1. A plot of the asymmetry index versus color for normal galaxies and galaxies undergoing mergers. The plot itself shows the normal systems which generally follow the sequence between (B-V) color and asymmetry, such that higher asymmetric galaxies have a bluer color. The galaxies which deviate from this relationship are those undergoing major mergers. This can be seen through the images of the labeled starbursts in the upper panel, which are all systems undergoing some type of major merger. Reproduced from Conselice et al. (2000a, 2000b).

robust non-parametric structural parameters. The formula for calculating the asymmetry index (A) is given by (Conselice 2003):

$$A = \min\left(\frac{\Sigma|I_0 - I_{180}|}{\Sigma|I_0|}\right) - \min\left(\frac{\Sigma|B_0 - B_{180}|}{\Sigma|I_0|}\right), \quad (3.5)$$

whereas in this equation the value I_0 represents the original galaxy image, and I_{180} is this image after rotating it by 180 deg from its center. Other rotational angles are possible and discussed in Conselice et al. (2000a). Within this calculation an iterative approach is used to find the center of rotation which gives the minimum asymmetry value. It is this value of the minimum asymmetry that is used as the final asymmetry value (e.g., Conselice et al. 2000a). The search radius to find the minimum asymmetry is typically a pixel or half a pixel, especially for HST imaging. Ultimately, as long as the search radius for the minimum is no smaller than about 1/300th of the galaxy's size, this should be

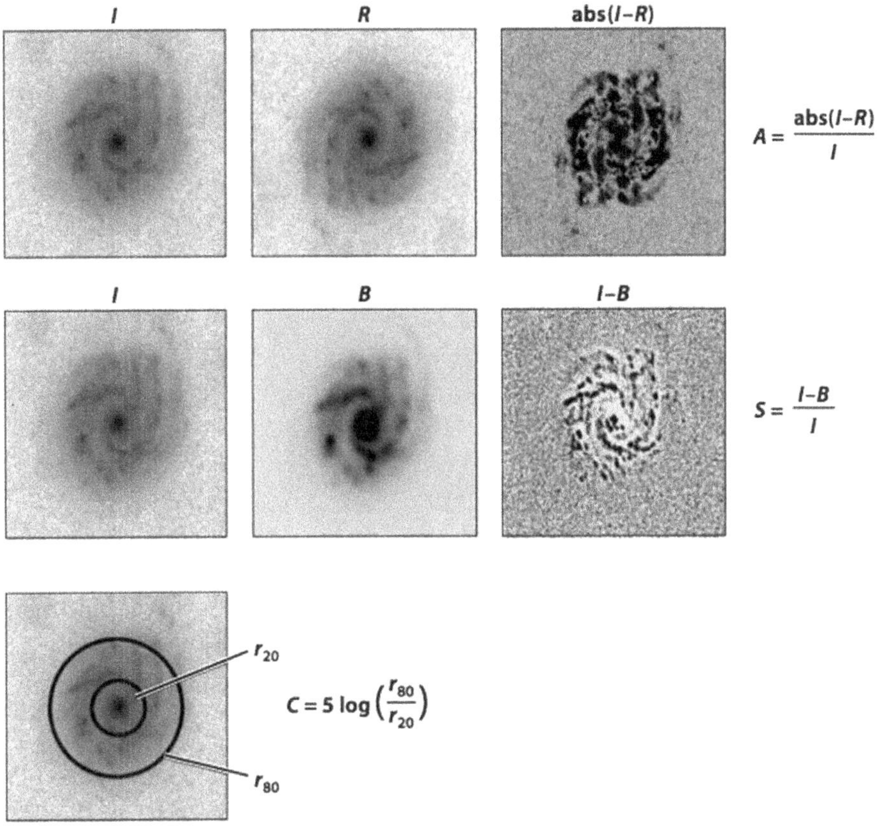

Figure 3.2. Plots showing the CAS method of quantifying galaxy structures. The computation of the asymmetry index (A) is shown in the top panel, whereby an example galaxy has been rotated (R) and subtracted from its original (I). The absolute values of the residuals are then summed and divided by the original flux of the galaxy to measure the asymmetry parameter. The clumpiness parameter (S) is in the next panel down and the concentration (C) is shown at the bottom. This shows the basic methodology for measuring the CAS parameters for galaxies. Note that there are many details in these measurements which are explained throughout the text of this chapter.

sufficient. For larger galaxies imaged at high resolution, such as nearby galaxies with the HST, this search radius should be larger, otherwise a local minimum will be obtained.

The values for the background *B* is computed in a similar way as for the galaxy itself, and this term is subtracted from the rotated and subtracted value for the galaxy image. Essentially, a single random area of the "sky" with no galaxies in it is used to determine this. This sky patch does not have to be the same size as that used in the galaxy measurement, as it can be normalized to match the size of the galaxy. It is usually smaller in size in fact. There are issues with background light being variable and newer methods use the median of all the asymmetry minimizations of the background within an image (Ferreira et al. 2020, in preparation). Note also that the measurement of the asymmetry index must be done within a consistent radius as asymmetry can vary with radii (Conselice et al. 2000a). The radius often used is the Petrosian radii discussed in Section 3.2.

The rotation and subtraction is done the same way in this "blank" area of sky as for the galaxy itself, the absolute value of this quantity is then taken and summed, and this "residual" is subtracted from the original asymmetry. This method ensures that for very faint and symmetric galaxies the asymmetry value is close to zero as it should be. Average values of the asymmetry index for different nearby galaxy types are listed in Table 3.1, showing that later-type galaxies are more likely to have a higher asymmetry, and that merging galaxies have by far the highest asymmetry values.

To demonstrate the basic ability of this, consider Figure 3.1, taken from Conselice et al. (2000a), which displays the color-asymmetry diagram for galaxies. The normal galaxies are found in the "main sequence" of color versus asymmetry, such that more asymmetric normal galaxies have a bluer color. The systems which deviate from this correlation are mergers—usually major mergers. The top part of Figure 3.1 shows images of these systems, highlighting the ones with circled crosses on the figure, which are the Starburst galaxies. This color-asymmetry diagram was the first combined use of a structural parameter and a physical one that correlates with the star formation rate and stellar populations (e.g., Conselice 1997). It can be used to find merging galaxies as those which have a range of colors, but very high asymmetries. A mixture of parameters with asymmetry can be used in the same way to find systems that are apparently undergoing merging—most commonly asymmetry plotted with the clumpiness or with the concentration.

Merging galaxies were examined in detail in Conselice et al. (2000a, 2000b) where it was shown that nearby merging systems deviate from scalings with asymmetry compared with normal galaxies. This is such that they have a higher asymmetry index at a given value of a parameter such as color or light concentration, as described earlier (see next section). This was later expanded in Conselice (2003) who furthermore showed that a new index call the clumpiness or smoothness index (S) can be used to find galaxies undergoing intense star formation, which can be used to furthermore find systems undergoing an active stage of merging while removing those systems with a distorted structure produced by intense star formation.

Table 3.1. The Average Concentration (C), Asymmetry (A), and Clumpiness (S) Parameters for Nearby Galaxies as Measured in the Optical R-Band (Reproduced from Conselice 2003.).

Galaxy Type	Concentration (R)	Asymmetry (R)	Clumpiness (R)
Ellipticals	4.4 ± 0.3	0.02 ± 0.02	0.00 ± 0.04
Early-type disks (Sa-Sb)	3.9 ± 0.5	0.07 ± 0.04	0.08 ± 0.08
Late-type disks (Sc-Sd)	3.1 ± 0.4	0.15 ± 0.06	0.29 ± 0.13
Irregulars	2.9 ± 0.3	0.17 ± 0.10	0.40 ± 0.20
Edge-on Disks	3.7 ± 0.6	0.17 ± 0.11	0.45 ± 0.20
ULIRGs	3.5 ± 0.7	0.32 ± 0.19	0.50 ± 0.40
Starbursts	2.7 ± 0.2	0.53 ± 0.22	0.74 ± 0.25
Dwarf Ellipticals	2.5 ± 0.3	0.02 ± 0.03	0.00 ± 0.06

3.3.2 Light Concentration

The examination of galaxy structure in a quantitative way was later expanded to robustly measure the concentration index. Within the CAS system, the first instance of this was in Bershady et al. (2000) who used the Petrosian radius as a new way to measure the total amount of light for deriving this parameter. The concentration of light is used as a method for quantifying the relative amount of light in the center of a galaxy compared to its outer parts. This concentration index C is in a sense a simple index, and correlates strongly with Sérsic n values, which are also a measure of the light concentration in a galaxy. It turns out that there are many ways of measuring the concentration of a galaxy, and these include ratios of radii which contain a certain fraction of light, as well as the ratio of the amount of light at two given radii (e.g., Bershady et al. 2000; Graham et al. 2005). The ratio of fluxes at different radii is however not a very robust way of measuring this index as it can change dramatically as a function of distance and redshift (e.g., Graham et al. 2005). A much more robust way to measure concentration is to take the ratio of radii which contain some fraction of the light within a galaxy.

The radii used for this are defined by the total amount of light measured within some Petrosian radius, often at the same location as used for measuring the asymmetry index. If the total amount of light differs from this then the resulting concentration index would also be different. Therefore, unlike other parameters, the exact definition of the total amount of light is critical for most measures of light concentration.

The definition most commonly used is the ratio of two circular radii which contain an inner and an outer fraction (20% and 80% or 30% and 70% are the most common) ($r_{\rm inner}$, $r_{\rm outer}$) of the total galaxy flux,

$$C = 5 \times \log\left(\frac{r_{\rm outer}}{r_{\rm inner}}\right). \tag{3.6}$$

Larger values of C reveal that larger fractions of light from a galaxy are contained within a central region. The concentration index, like the asymmetry index, must be measured very carefully, as different regions and radii used can produce very different values that systematically do not reproduce well when observed under different conditions (e.g., Graham et al. 2001; Conselice 2003; Graham et al. 2005). The values of the concentration parameters for typical galaxies are shown in Table 3.1.

Concentration can also be used to separate galaxies with different star formation histories in the local universe. By using SDSS data, Strateva et al. (2001) demonstrated that non-star-forming galaxies are more concentrated than star-forming blue galaxies. This type of correlation with concentration is also seen with other galaxy properties (e.g., Allen et al. 2006; Conselice 2006a). There is also a link between the central massive black holes at the centers of these galaxies and their light concentrations. Within at least ellipticals the light concentration correlates with the mass of the central massive black hole (e.g., Savorgnan et al. 2013). What is producing this correlation is not understood, and it might be the result of an overall correlation between concentration and mass which then determines the mass of the central black hole.

We have now discussed the asymmetry and concentration parameters, which are the two, historically speaking, primary parameters for describing galaxy properties in a non-parametric way. We can now compare these two parameters and determine where nearby galaxy populations fall within various parameter spaces. One example of this is shown in Figure 3.3. As can be seen here, different galaxy types fall in different parts of this parameter space. The ellipticals or early types are found in the higher concentration and low asymmetry regions. Disk galaxies have modest concentrations around values of $C \sim 2.5$–3, and modest asymmetries. Peculiar galaxies have a range of concentration values, but generally have large asymmetries.

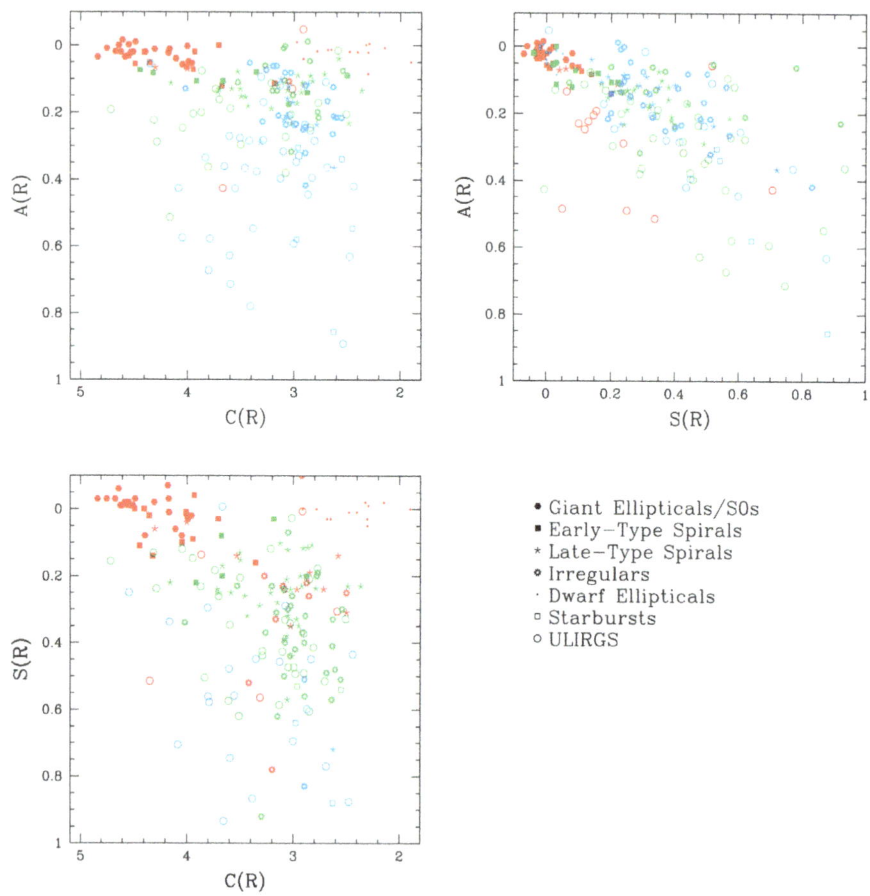

Figure 3.3. Plots showing the use of quantifying galaxy structures to find different galaxy types with the CAS parameters. Each parameter is plotted against the other in these three panels. The plot at the top left shows where galaxy types seen in the legend fall in the asymmetry-concentration diagram. As can be seen, ULIRGs, which are often major mergers, have a uniquely high asymmetry value. Other galaxy types fall in different parts of this parameter space, as can be seen. Reproduced from Conselice (2003).

There is however quite a large spread of values for each type and overlaps in galaxy types within CAS parameter space as well. Therefore these types of diagrams can be used to find galaxies of different types, but they must be used with caution given the overlapping values of these parameters for different galaxy types. This issue will be revisited several times throughout this book and in particular in Chapter 8 when we discuss machine learning applications to morphology and structure.

3.3.3 Clumpiness

We describe in the section how the clumpiness (or smoothness) (S) parameter is calculated. This is the third parameter of the CAS parameters and the last to be introduced in Conselice (2003). In many ways, it is still a largely unexplored parameter. This parameter is used to characterize how much light in a galaxy is made up of clumpy structures, such as star-forming regions. A galaxy which is completely smooth with no "modeling" will have an $S \sim 0$ value. On the other hand, galaxies with many regions or clumps that contain a large fraction of the light in the galaxy will have a high S value. Star-forming galaxies tend to have many clumpy structures, resulting in high S values. In fact, the clumpiness is found to correlate with a measure of the star formation rate (e.g., Conselice 2003). The clumpiness is defined as described in Conselice (2003) as:

$$S = 10 \times \left[\left(\frac{\Sigma(I_{x,y} - I_{x,y}^{\sigma})}{\Sigma I_{x,y}}\right) - \left(\frac{\Sigma(B_{x,y} - B_{x,y}^{\sigma})}{\Sigma I_{x,y}}\right)\right], \tag{3.7}$$

whereby the original image $I_{x,y}$ is blurred to produce a secondary image, $I_{x,y}^{\sigma}$. The image $I_{x,y}^{\sigma}$ is produced by convolving the original image with a Gaussian kernel of some FWHM which is determined by the size of the galaxy itself. This is in a sense a way to "blur" the galaxy. This blurred image is then subtracted from the original image which produces a new image or residual map. This residual image of the smoothed image minus the original will only contain high-frequency structures in the galaxy (Conselice 2003). For computational purposes, the size of the smoothing kernel σ is determined by the radius of the galaxy. From experimentation it is found that the value $\sigma = 0.2 \cdot 1.5 \times r(\eta = 0.2)$ provides the best signal for nearby systems (Conselice 2003). There are other features of the computation of S that must also be considered when carrying out this computation. The centers of galaxies are removed when this procedure is carried out as they contain unresolved high-spatial frequency light. This is often the inner 5%–10% of the inner pixels within a galaxy which are set to 0 within the calculation of the S parameter.

3.3.4 Other Coefficients

Other structural measurements include the Gini/M_{20} parameters, which are used in a similar way to the CAS parameters to classify galaxies into broad morphological types, especially those undergoing mergers (e.g., Abraham et al. 2003; Lotz et al. 2004). These parameters measure the relative distribution of light within pixels, and have one benefit in that they do not involve subtraction, as is used for the asymmetry and clumpiness parameters. Because of this, they in principle are less affected by noise for lower S/N pixels. Thus, in principle, these two parameters may be less sensitive to high levels of background noise (e.g., Lotz et al. 2004). However, the radius used to define these parameters can make a major difference in their values. It is also the case that distances may have a larger effect on these parameters than the CAS parameters.

The Gini index was originally developed by the Italian sociologist and statistician Corrado Gini who invented it as a way to account for the differences in the distribution of wealth within nations. The Gini value is generally defined as having a range that goes from 0 (complete equality in wealth) to 1 (complete inequality, i.e., one "king" who owns everything). The Gini index is most commonly used in economics, and is still used to describe the inequality of nations compared with each other. South Africa has the highest Gini index (in 2020) of 0.63, and the lowest is Ukraine with a value of 0.25. Two galaxies which are within the range of Gini values for nations are shown in Figure 3.4.

The value of the Gini value, G, is defined by the Lorentz curve of the galaxy's light distribution. This curve has no spatial position information considered in its calculation. However, the radius is important as any pixels included within the calculation will alter the value of the Gini index. Thus, measuring a galaxy with a larger radius will include more pixels, giving a larger Gini index as most of these pixels will not have much light in them.

To calculate these parameters the pixels in a galaxy are ordered by their brightness and added together as part of the cumulative distribution of light (see Lotz et al. 2004). A galaxy in this framework, considered as an imaged system with n pixels, with each pixel i containing a flux f_i where i ranges from 0 to n, the total number of pixels in the galaxy.

The Gini coefficient calculation for a galaxy is then given by:

$$G = \frac{1}{|\bar{f}|n(n-1)} \sum_{i}^{n} (2i - n - 1)|f_i|, \tag{3.8}$$

Figure 3.4. The Gini index for two galaxies which span the range of the Lotz et al. (2004) measurements of the Frei et al. (1996) catalog. The value of the Gini index for NGC 3377 is similar to that of the most unequal countries, including South Africa, whereas NGC 4571 has a lower Gini index similar to the distribution of wealth in the most equal countries. Courtesy of NASA.

where the value $\bar{f}$ is the average pixel flux value. Typically the Gini coefficient ranges from 0 to 1, whereby the higher the Gini index the higher fraction of the light within a galaxy originates from fewer pixels.

The second-order moment parameter, M_{20}, is similar to the concentration as it indicates how much light is concentrated within an image. It differs somewhat from the concentration, as a lower M_{20} value, which assumes a high "concentration" when more negative, does not necessarily imply a central concentration. In principle, this concentration could arise from a collection of light at any location in a galaxy. That is, M_{20} gives us some idea of the fraction of light which is concentrated anywhere within a galaxy, whereas the concentration parameter gives an indication of how centrally concentrated the light in a galaxy is. Usually, these two parameters will correlate, but when they do not it usually indicates that there is a concentration of light within the galaxy that is not located at the center of a galaxy.

Mathematically, the value of M_{20} is the moment of the fluxes of the brightest 20% of light in a galaxy, normalized by the total light moment of all the pixels in a galaxy (Lotz et al. 2004, 2008). The calculation for the M_{20} index is done through the equation:

$$M_{20} = \log 10\left(\frac{\sum_{i}}{M_{\text{tot}}}\right) \text{while} \sum_{i} f_i < 0.2 f_{\text{tot}}, \tag{3.9}$$

where the value of M_{tot} is

$$M_{\text{tot}} = \sum_{i}^{n} M_i = \sum_{i}^{n} f_i\left[(x_i - x_c)^2 + (y_i - y_c)^2\right],$$

where the values x_c and y_c are the pixel locations of the center of a galaxy. In the case of the M_{20} values, this center is defined as the location where the value of M_{tot} is minimized (Lotz et al. 2004), similar to how the asymmetry is calculated. The separation ability for distinguishing nearby elliptical, spirals and ULIRGs is similar in purity and completeness to that found for the CAS parameters (see Lotz et al. 2004).

Other popular parameters include the multiplicity index, Ψ, which is a measure of the potential energy of a light distribution (e.g., Law et al. 2007). Ψ values can range from 0 for galaxies in a compact form, to values $\Psi > 10$ that are found for irregular/peculiars (e.g., Law et al. 2012). Other studies by Freeman et al. (2013) include measures such as parameters that measure multi-mode (M), intensity (I), and deviation (D) of a galaxy's light profile. These are designed especially for finding galaxies undergoing mergers, which is one of the most powerful things these indices can do.

3.4 Identifying Mergers

One of the main uses of galaxy structure is to identify unusual modes of galaxy formation. The most important of these in terms of mass assembly is the merger process. Merging is also a key prediction of cold dark matter models, providing the dominant idea for how galaxy evolution occurs through cosmic time (e.g., White & Rees 1978; Blumenthal et al. 1984; Cole et al. 2000). The most common type of mergers studied to date are major mergers—merging systems where the mass ratio of the two galaxies merging goes from the same mass (1:1) down to (1:3) or (1:4) mass ratios. Major mergers dominate the study of merging as they are often the easiest type to find and examine and are relatively easy to simulate on computers. The reason for this is that for ongoing major mergers the effects on the galaxy are easiest to see, as similar mass galaxies produce a maximum level of visual "destructive" effects on the galaxies during a major merger.

The other main type of galaxy mergers are the so-called minor mergers, whereby the mass ratio is lower than (1:3). Minor mergers often do not produce a significant morphological or structural effect which has the downside of making them difficult to locate and study. However, they are predicted and observed to be more common than major mergers and may turn out to be more important for the galaxy formation and evolution process. Ultimately, minor mergers can be studied with structures, but care must be taken as contamination from non-merger sources is more likely to produce a signal than a real minor merger. Yet another merger type, almost unstudied to date due to the fact that it is difficult to find them, are the micro-mergers which occur when very low mass galaxies with masses <1/100th merge with a large mass system. These micro-mergers likely are responsible for the creation of extensive amounts of cold gas, which can induce intense star formation and therefore galaxy building, a topic we discuss in Chapter 6.

In general, there are two main types of ways to find galaxies involved in mergers: the pair method and through analyzing structures. Galaxy mergers are found in pairs when two galaxies of a given mass ratio are separated by a certain distance (see Figure 3.5). The separation used to find these pairs is almost always 20 kpc or 30 kpc within the published literature (e.g., Mundy et al. 2017; Duncan et al. 2019). For kinematic pairs the selection for mergers is that the two galaxies be close together on the sky in projection, but to also have a low velocity difference of around 200 km s^{-1}. Larger separations are also used, however these galaxies are unlikely to merge within a reasonable timescale and are a measure of galaxy clustering on larger scales. Two galaxies that are as close as <30 kpc have a good chance of merging within a calculated dynamical friction timescale (see Chapter 4) that is observable. The timescale furthermore allows us to convert the merger fraction into a merger rate as we know the timescale in which these systems will remain in the state they are observed in. Pairs have the benefit that we know that the two galaxies are close together and will therefore possibly merge, but there is no guarantee that they will do so. To know for sure that a galaxy has merged we need to examine those systems that have already merged—and that is best done through galaxy structural analyses.

Identifying galaxies merging with structure is the other method for finding mergers, which requires high-resolution imaging, such as with the HST for distant galaxies (see Chapter 4), or nearby galaxy imaging. Finding structural mergers is a way to measure the later part of the merger process when the two separate galaxies are coalescing into a single system. The physics behind this merging, and the later formation into a new system, has been modeled through the dynamical process

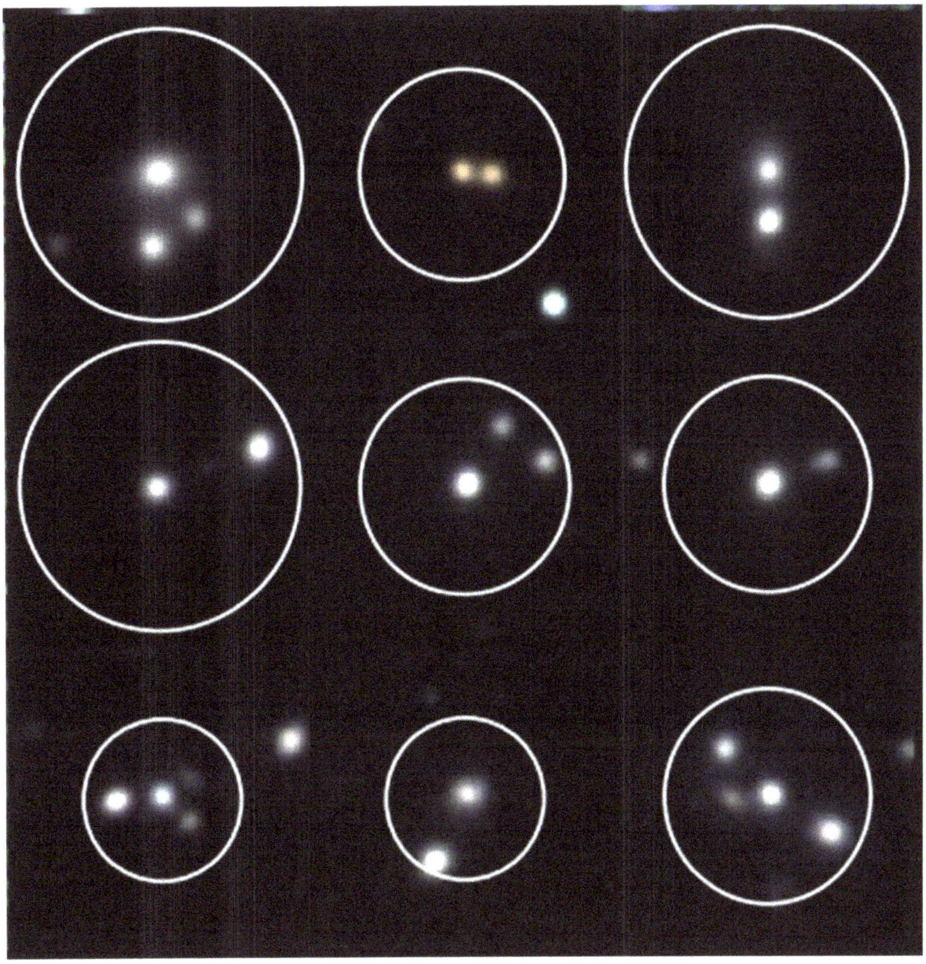

Figure 3.5. Shown are examples of galaxy pairs as seen in the distant universe from Mundy et al. (2017). Each of the circles shown contains at least two different galaxies as separate systems. However, due to calculations of merger timescales, these systems are close enough such that dynamical friction will make them merge within a few billion years. These particular real examples are major mergers, such that the mass of the smaller galaxy is at least 25% of the mass of the larger system.

called dynamical violent relaxation (e.g., Lynden-Bell 1967). Galaxies undergoing these later stages of the merging process will be seen by eye as peculiar galaxies, such as those seen by Arp (1966). Examples of these peculiar systems are shown in Chapter 1. It is also possible that some of the peculiars are peculiar not because of a merger but because of another cause, such as extensive star formation or galactic outflows. We later discuss how to separate the actual peculiars which are mergers from these systems.

Both of the methods described above for finding mergers have their disadvantages and distinct advantages, some of which have already been described. When using pairs of galaxies to find mergers we do not know for certain whether the two galaxies are in the same physical aspect of space, as there can be chance superpositions of galaxies at different distances that appear to be close together in the sky, but in actuality are very distant in the line of sight direction. This can be alleviated by photometric, or in unique cases spectroscopic redshifts (e.g., Tasca et al. 2014; Mundy et al. 2017). For those galaxies undergoing a merger and are peculiar in appearance our ultimate aim is to distinguish mergers from systems which are undergoing significant star formation, which can make a galaxy appear to be peculiar in appearance, particularly at high redshifts.

The first attempts to find mergers were largely based on discovering galaxies in pairs that originated from the goal of deriving cosmological parameters from galaxy mergers (e.g., Patton et al. 1997). Later these studies were extended out to slightly higher redshifts by, e.g., Le Fèvre et al. (2000) and López-Sanjuan et al. (2011, 2012). Pair studies can now be carried out to $z \sim 6$ (e.g., Duncan et al. 2019). This work is described in detail in Chapter 5.

The morphological method for finding mergers really began to be used in the mid-1990s when the Hubble Deep Fields programs were initiated. It is not however just simply enough to find galaxies that are peculiar to identify unambiguously a merger, but to use quantitative ways to define and locate merging systems. From this, we can statistically determine whether or not the system is a merger. In contrast, very early studies used visual morphologies to determine whether or not a galaxy was undergoing a merger (e.g., Le Fèvre et al. 2000).

However, one of the largest problems with using structure to find galaxies that are merging in the distant universe is that images of distant galaxies are probing wavelengths that are much bluer than the typical wavelengths used to measure galaxy structure in the nearby universe that are typically in rest-frame optical light. This is called the morphological k-correction, which can be a significant effect in the quantitative structure of the system (Mager et al. 2018). This is because we are seeing younger stars when we probe bluer light, and the distribution of these stars can be very different from what appears in redder light that originates from older more evolved stars.

The HST first examined distant galaxies using the Wide-Field Planetary Camera (WFPC) and WFPC2 cameras. These were optical cameras whose reddest filter goes only to about 9000Å. When these cameras were used to study galaxies at redshifts at $z > 1.5$ we could only observe their structures in their rest-frame ultraviolet morphologies. In 1998 the HST was refurbished when a shuttle servicing mission installed a new camera called the Near Infrared Camera and Multi-Object Spectrometer (NICMOS) allowing us to probe into the rest-frame optical for galaxies up to $z \sim 2.5$.

NICMOS was the first Hubble camera that allowed us to image the rest-frame optical light of distant galaxies at high resolution. Because young stars can dominate the morphologies of galaxies in the ultraviolet, it was not understood whether the peculiar galaxies seen with the optical cameras on Hubble were simply just due to the fact that we were looking at the star formation within the galaxy in the ultraviolet or to an underlying disturbance due to a merger that has created a peculiar galaxy.

Not long after the NICMOS camera was installed an infrared version of the Hubble Deep Field was obtained (Dickinson et al. 2000). This new imaging showed that the structures and morphologies of distant galaxies were very similar in the rest-frame UV compared with rest-frame optical (Papovich et al. 2003, 2005; Conselice et al. 2005). This is often the case when a system is dominated by star formation; its morphology is similar at all wavelengths, as can be seen when viewing nearby galaxies at UV and optical wavelengths (e.g., Windhorst et al. 2002).

The clumpiness and the asymmetry index can be used together to find systems that deviate from a tight linear relationship. This deviation is such that mergers have a high asymmetry but a modest clumpiness, or concentration (e.g., Figure 3.6). The idea behind this deviation is that if a galaxy has a high clumpiness it implies a significant fraction of light from that galaxy originates from small-scale structures in the galaxy, typically star-forming knots. Light from galaxy mergers does not produce such small-scale structures, but bulk large-scale asymmetric ones. In this sense, an intensely star-forming galaxy will have a resulting high asymmetry and a high clumpiness. However, a merging system usually produces a high asymmetry without a corresponding high clumpiness which would be produced in an intense star formation episode. This is a very effective way in which to distinguish star-forming galaxies from those undergoing mergers (e.g., Conselice 2003).

This C versus A methodology works well for locating mergers in the nearby universe. This can be shown by the systematic deviation of ultra-luminous infrared galaxies (ULIRGs) in the asymmetry and clumpiness plane from normal galaxy such as spirals and ellipticals. What is found empirically is that mergers have an asymmetry value A greater than $A > 0.35$ for a major merger. To remove galaxies undergoing intense star formation from actual mergers the criteria $A > S$ must also

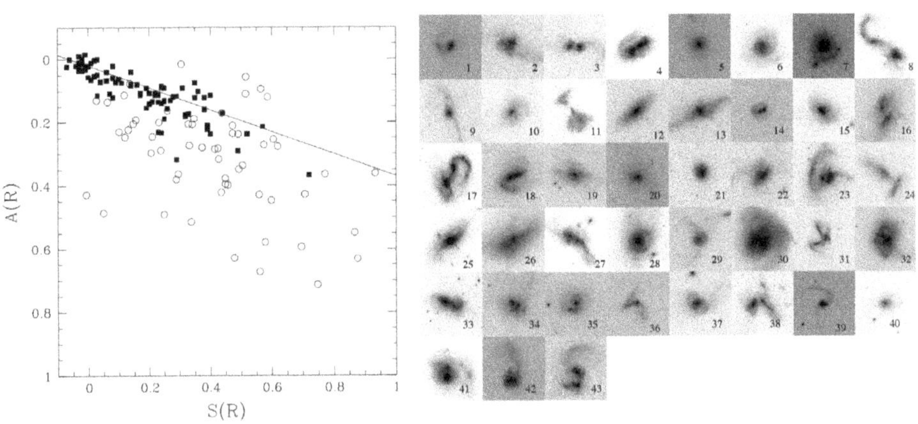

Figure 3.6. On the left: the asymmetry parameter versus the clumpiness index, plotted such that the normal galaxies that are not merging are plotted as black squares, and the systems undergoing mergers are shown as the open circles. These clearly deviate with a higher asymmetry at a given clumpiness. On the right-hand side is shown the images of the galaxies which are undergoing mergers which make up the open points on the left. These are ultra-luminous infrared galaxies (ULIRGs) which have been shown to be dominated by the merger process.

hold (Conselice 2003) to select mergers. Overall the following criteria must hold to select a major merger by structure:

$$A > 0.35\&A > S. \tag{3.10}$$

When this methodology is applied to distant galaxies what we find is that the derived merger fraction increases significantly when we go to higher redshifts (Chapter 5). The merger fraction was first studied in this way in Conselice et al. (2003) who showed that the peak merger fraction at $z = 2.5$ was about 40% using a method such as this, which has been confirmed in numerous other studies since.

This formalism accounts for most, but not all, of local mergers—i.e., ULIRGs and other mergers. The contamination or false-positive fraction using the CAS method by non-mergers is fairly low at a few percent. However, one issue is that the fraction of actual mergers in the defined selection using CAS identified mergers is roughly 50% (Conselice 2006b). This is due to galaxies involved in the merger process only being quantitatively asymmetric, as measured by the asymmetry index, for about a third of the life-time of the merger (see Chapter 4 for a discussion of computer simulations of morphology). CAS parameters are not the only way to find galaxy mergers with structure. There also exists the method explained by Lotz et al. (2004) for locating mergers using the Gini/M_{20} parameters. The criteria for selecting mergers with these parameters is given by,

$$G > -0.14 \times M_{20} + 0.33. \tag{3.11}$$

Another merger criteria developed by Freeman et al. (2013) uses multi-mode (M), intensity (I), and deviation (D) statistics to find which galaxies are mergers. A higher fraction of real mergers are found using these indices compared with CAS or Gini/M_{20}. Most recently it appears that machine learning techniques may supersede all of these methods for finding galaxies in mergers, which appear to already be quite successful (e.g., Snyder et al. 2019; Ferreira et al. 2020; Chapter 8).

3.5 Non-Evolutionary Effects on Structure

Below we discuss some of the biases, or non-evolutionary effects, which can mimic real evolution when examining structural parameters at different redshifts and distances. This is an important thing to correct for, because if we blindly apply the same methods to low and high redshifts we will see an apparent change in structure, but that does not imply that real evolution has occurred. Since structural evolution of galaxies is the main aspect of this book, it is critical to try to understand these issues in some detail. These issues are also explored in alternative and more detailed ways in Chapters 4 and 5.

It is thus important to sort out these effects of distance and redshift before any real evolution can be inferred. The two methods described here are the signal to noise (S/N) and resolution effects produced by distance, and the change in observed wavelength due to redshift, which we call the morphological k-correction. Ultimately both effects are corrected before any real changes in galaxies can be claimed.

3.5.1 Image Simulations—Effects of Redshift on Structure

A major important issue with non-parametric structural index measurements, as well as those that are parametric, is that the measured or fitted values will change for more distant galaxies, due to any real evolution, but also because of distance effects. The effects of distance create an observed smaller and fainter image of the same system. At the same time, the surface brightness of the object will decline at higher redshifts as $\sim(1 + z)^{-4}$ (Whitney et al. 2019). It is critical to account for these effects when using quantitative or even qualitative galaxy structure as a measure of evolution (e.g., Conselice 2003; Lisker 2008; Conselice et al. 2011).

There are however several methods to correct this issue. In many ways this correction is similar to how corrections for point spread functions in parametric fitting or weak lensing analyses are done. The most common method for correcting for distances in these non-parametric parameters, and even for parametric fitting (e.g., Buitrago et al. 2008) is to use image simulations to determine the effects of distance on the measurement of structure. The simulations needed to make this correction are such that nearby galaxies are reduced in resolution and surface brightness to match the values that would be obtained at the redshift at which the galaxy is to be simulated to. These new simulated images are then placed into an accurate background of the image in question, which depends on the telescope and the depth of the imaging and the filter. For example, we know how the background should look in any HST image in any filter at a given depth. This background is critical and depends on the instrument and exposure time in which the simulation is trying to mimic (Conselice 2003). Another method is to actually place the simulated image into existing data in a random location where no previous galaxy exists. The outline for these simulations is discussed in papers such as Giavalisco et al. (1996), Conselice (2003), Whitney et al. (2019), among others.

As an example of the difficultly in reproducing the morphologies and structures of galaxies at different redshifts, Figure 3.7 shows simulated nearby early-type spirals as to how they would appear in WFC3 imaging taken by the Hubble Ultra Deep Field (Conselice et al. 2011). It can be clearly seen that it is difficult to make out features of these simulated galaxies after they have been simulated, and at times the overall morphology as obtained by eye is very different. In this case, these nearby spirals appear to be more elliptical or early-type, than they were before the simulation. There would also be a change in the morphological measurements, such that these galaxies become less asymmetric and more concentrated. If this is not taken into account when evolution is derived, then a false measurement will be obtained.

The process for accounting for image degradation due to redshift is through the image simulations described above by moving a galaxy from low to high redshift. After this is done the morphological index of interest is measured at $z = 0$, in its "ground state" and then the same parameter is remeasured within the image at higher redshifts after simulating.

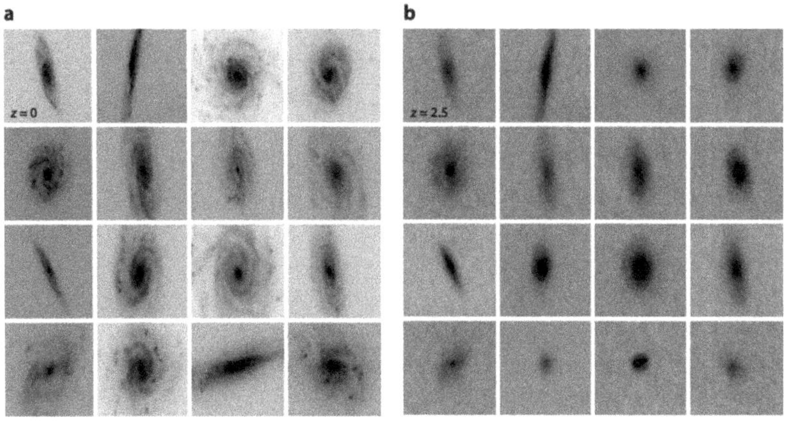

Figure 3.7. Image simulations of galaxies from low redshift to high redshift as seen in deep HST imaging. Nearby galaxies originally observed at $z \sim 0$ in the rest-frame B-band are shown on the left panel (a). These same galaxies after being simulated to how they would appear at $z = 2.5$ are shown on the right panel (b). In this simulation the simulated galaxies appear as if they were observed in the rest-frame B-band, within the Hubble Ultra Deep Field WFC3 F160W (H) band. These systems are classified as late-type spirals (Sc and Sd) in the nearby universe, but can appear very different when simulated to higher redshifts as seen here, and when using quantitative measures. Typically, the galaxies look more early-type and more concentrated. The typical sizes of these galaxies are several kpc in effective radii, and are at a variety of distances (see Conselice et al. 2000a, 2011). These changes in structure, both in apparent morphology and in terms of the structural indices must be carefully considered before evolution is derived by comparing galaxies at different redshifts (e.g., Conselice et al. 2008; Mortlock et al. 2013).

3.5.2 Quantifying Morphological k-Corrections

As already described, another issue when examining distant galaxies is that these systems will often be imaged at bluer wavelengths than typically observed in the nearby universe due to the effects of redshift. The qualitative and quantitative structures of galaxies can often change significantly between rest-frame ultraviolet and rest-frame optical images (e.g., Meurer et al. 1995; Hibbard & Vacca 1997; Windhorst et al. 2002; Taylor-Mager et al. 2007). Images of galaxies at $z > 1.2$ taken with WFPC2 and ACS are all imaged in the rest-frame ultraviolet, which may present a problem with interpreting their structures.

Figure 3.8 shows the rest-frame wavelengths of various popular HST filters and the observed wavelengths each probe as a function of redshift. This demonstrates that observations must be done in the near-infrared to probe rest-frame optical light for galaxies at $z > 1$. Due to the fact that when galaxies are undergoing intense star formation their morphologies in the rest-frame optical and rest-frame UV are very similar, such that their structures can give very similar visual and qualitatively based morphologies (Dickinson et al. 2000). This is however not true for galaxies that have a large amount of dust, as these galaxies can be quite dust extincted in the UV, and thus their structures would be much fainter and harder to distinguish. However, these dusty galaxies are relatively rare, and for the most part are not considered extensively in this book. Ultimately what is done to correct for the effects of the

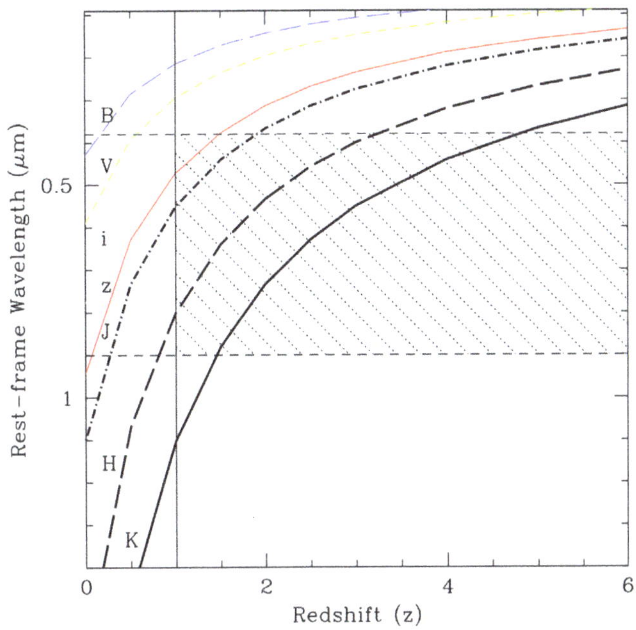

Figure 3.8. The change of observed wavelength with redshift for galaxies observed with typical Hubble Space Telescope filters, as a function of redshift from $z \sim 0 - 6$. The filters shown are: ACS B_{450}, V_{550}, i_{775}, z_{950} filters, and the WFC3 J_{110} and H_{160} filters and a K-band filter centered at 2.2 μm. The shaded area shows the region in which the rest-frame optical light of distant galaxies can be probed from between 0.38 μm to 0.9 μm. These are the typical wavelengths in which most galaxy studies in the local universe have occurred in. As can be seen, the H-band allows rest-frame light up to $z \sim 3$ to be imaged, while the K-band (which HST does not have) can extend this out to $z \sim 4.5$.

morphological k-correction is to have some understanding of how the quantitative structure as measured in the CAS parameters, for example, changes between two different wavelengths and then to apply this correction when observing a galaxy at a bluer wavelength where telescopes can often probe, to redder wavelengths where these morphological and structural systems are calibrated. The limitation to this method of correcting for morphological k-corrections is that the data to carry out this correction is often at a different redshift than the one being corrected, and there is no guarantee that the properties of the quantitative structures of these galaxies at two different redshifts will be the same.

The approach for correcting for the morphological k-correction is to measure the parameter of interest at different wavelengths, and determine by interpolation the value at the rest-frame wavelength of interest, which is usually optical light. To consider how this work in practice we give an example of how the correction is done for the asymmetry index (A).

In this example the observed asymmetry for a galaxy at some redshift z is $A_{\rm obs}$. However, because of the effects of redshift, there is a correction for redshift effects (called SB dimming; described in Section 3.5.1) and for the k-correction, discussed above. When we consider these terms the final actual measurement value of the asymmetry can be written as:

$$A_{\text{final}} = (A_{\text{obs}} + \delta A_{\text{SB-dim}} + \delta A_{\text{k-corr}}). \tag{3.12}$$

The value $\delta A_{\text{k-corr}}$ is the (usually negative value, decreasing the measured asymmetry) morphological k-correction, while $\delta A_{\text{SB-dim}}$ is the (positive) correction for image degradation effects. Corrections for other parameters can be done in a similar way. Note that the k-correction term is only applicable if the rest-frame observed wavelength differs from the rest-frame wavelength desired for the parameter being measured. This correction is necessary to obtain parameters that can be compared between redshifts to determine galaxy populations and how they evolve. If the rest-frame wavelength is the same throughout a structural evolution analysis, then the term $A_{\text{k-corr}}$ is set equal to zero.

3.6 Summary

This chapter is a detailed one on the methods for measuring the quantitative structures of galaxies. We have introduced parametric fitting, such as through Sérsic profiles and fits, but also non-parametric measurements that do not assume a form, such as the CAS parameters. This chapter does not apply these to galaxies (see Chapter 5 for this), but explains how they are measured. A reader interested in developing these methods will be able to reproduce these measurement tools from the descriptions here or to use publicly available software to measure them on galaxies. This chapter also discusses the uses of these parameters to find different galaxy types, but primarily their usefulness is for locating galaxies undergoing merging. The next chapter introduces the high-redshift universe and begins the description of how these parameters are used to derive the structural evolution of galaxies up to $z \sim 6$.

References

Abraham, R. G., van den Bergh, S., Glazebrook, K., et al. 1996, ApJS, 107, 1

Abraham, R. G., van den Bergh, S., & Nair, P. 2003, ApJ, 588, 218

Allen, P. D., Driver, S. P., Graham, A. W., et al. 2006, MNRAS, 371, 2

Arp, H. 1966, ApJS, 14, 1

Bershady, M. A., Jangren, A., & Conselice, C. J. 2000, AJ, 119, 2645

Blumenthal, G. R., Faber, S. M., Primack, J. R., et al. 1984, Natur, 311, 517

Buitrago, F., Trujillo, I., Conselice, C. J., et al. 2008, ApJL, 687, L61

Buitrago, F., Trujillo, I., Conselice, C. J., et al. 2013, MNRAS, 428, 1460

Cole, S., Lacey, C. G., Baugh, C. M., et al. 2000, MNRAS, 319, 168

Conselice, C. J. 1997, PASP, 109, 1251

Conselice, C. J., Bershady, M. A., & Jangren, A. 2000a, ApJ, 529, 886

Conselice, C. J., Bershady, M. A., & Gallagher, J. S. 2000b, A&A, 354, 21L

Conselice, C. J. 2003, ApJS, 147, 1

Conselice, C. J., Bershady, M. A., Dickinson, M., & Papovich, C. 2003, AJ, 126, 1183

Conselice, C. J., Blackburne, J. A., & Papovich, C. 2005, ApJ, 620, 564

Conselice, C. J. 2006a, MNRAS, 373, 1389
Conselice, C. J. 2006b, ApJ, 638, 686
Conselice, C. J., Rajgor, S., & Myers, R. 2008, MNRAS, 386, 909
Conselice, C. J., Bluck, A. F. L., Buitrago, F., et al. 2011, MNRAS, 413, 80
de Vaucouleurs, G. 1948, AnAp, 11, 247
Dickinson, M., Hanley, C., Elston, R., et al. 2000, ApJ, 531, 624
Duncan, K., Conselice, C. J., Mundy, C., et al. 2019, ApJ, 876, 110
Erwin, P. 2015, ApJ, 799, 226
Ferreira, L., Conselice, C. J., Duncan, K., et al. 2020, ApJ, 895, 115
Freeman, P. E., Izbicki, R., Lee, A. B., et al. 2013, MNRAS, 434, 282
Frei, Z., Guhathakurta, P., Gunn, J. E., et al. 1996, AJ, 111, 174
Giavalisco, M., Steidel, C. C., & Macchetto, F. D. 1996, ApJ, 470, 189
Graham, A. W., Erwin, P., Caon, N., & Trujillo, I. 2001, ApJ, 563, 11
Graham, A. W., Driver, S. P., Petrosian, V., et al. 2005, AJ, 130, 1535
Hibbard, J., & Vacca, W. D. 1997, AJ, 114, 1741
Kormendy, J., Fisher, D. B., Cornell, M. E., & Bender, R. 2009, ApJS, 182, 216
Kormendy, J., & Djorgovski, S. 1989, ARA&A, 27, 235
Lange, R., Driver, S. P., Robotham, A. S. G., et al. 2015, MNRAS, 447, 2603
Law, D. R., Steidel, C. C., Erb, D. K., et al. 2007, ApJ, 656, 1
Law, D. R., Steidel, C. C., Shapley, A. E., et al. 2012, ApJ, 745, 85
Le Fèvre, O., Abraham, R., Lilly, S. J., et al. 2000, MNRAS, 311, 565
Lintott, C., Schawinski, K., Bamford, S., et al. 2011, MNRAS, 410, 166
Lisker, T. 2008, ApJS, 179, 319
López-Sanjuan, C., Le Fèvre, O., de Ravel, L., et al. 2011, A&A, 530, A20
López-Sanjuan, C., Le Fèvre, O., Ilbert, O., et al. 2012, A&A, 548, A7
Lotz, J. M., Primack, J., & Madau, P. 2004, AJ, 128, 163
Lotz, J. M., Davis, M., Faber, S. M., et al. 2008, ApJ, 672, 177
Lynden-Bell, D. 1967, MNRAS, 136, 101
Mager, V. A., Conselice, C. J., Seibert, M., et al. 2018, ApJ, 864, 123
Meurer, G. R., Heckman, T. M., Leitherer, C., et al. 1995, AJ, 110, 2665
Mortlock, A., Conselice, C. J., Hartley, W., et al. 2013, MNRAS, 433, 1185
Mundy, C. J., Conselice, C. J., Duncan, K. J., et al. 2017, MNRAS, 470, 3507
Papovich, C., Giavalisco, M., Dickinson, M., et al. 2003, ApJ, 598, 827
Papovich, C., Dickinson, M., Giavalisco, M., et al. 2005, ApJ, 631, 101
Peng, C. Y., Ho, L. C., Impey, C. D., & Rix, H.-W. 2002, AJ, 124, 266
Patton, D. R., Pritchet, C. J., Yee, H. K. C., et al. 1997, ApJ, 475, 29
Petrosian, V. 1976, ApJ, 209, 1L
Redman, R. O. 1936, MNRAS, 96, 588
Savorgnan, G., Graham, A. W., Marconi, A., et al. 2013, MNRAS, 434, 387
Schade, D., Lilly, S. J., Crampton, D., et al. 1995, ApJ, 451, 1L
Sérsic, J. L. 1963, BAAA, 6, 41
Simard, L., Meden, J. T., Patton, D. R., Ellison, S. L., & McConnachie, A. W. 2011, ApJS, 196, 11
Snyder, G. F., Rodriguez-Gomez, V., Lotz, J. M., et al. 2019, MNRAS, 486, 3702
Steinmetz, M., & Navarro, J. F. 2002, NewA, 7, 155
Strateva, I., Ivezić, Ž., Knapp, G. R., et al. 2001, AJ, 122, 1861
Tasca, L. A. M., Le Fèvre, O., López-Sanjuan, C., et al. 2014, A&A, 565, A10

Taylor-Mager, V., Conselice, C. J., Windhorst, R., & Jansen, R. 2007, ApJ, 659, 162
Trujillo, I., Conselice, C. J., Bundy, K., et al. 2007, MNRAS, 382, 109
White, S. D. M., & Rees, M. J. 1978, MNRAS, 183, 341
Whitney, A., Conselice, C. J., Bhatawdekar, R., et al. 2019, ApJ, 887, 113
Windhorst, R. A., Taylor, V. A., Jansen, R. A., et al. 2002, ApJS, 143, 113

Chapter 4

The Distant Universe

In this chapter, the high redshift universe is introduced. This includes a discussion of the Hubble Deep Field (HDF) which sets the standard for this type of analysis, and which largely put in motion the study of high redshift galaxy structure, starting in 1996 in earnest, although with some previous work. This chapter will also describe what we have learned about the sizes, stellar masses, and morphologies of galaxies observationally at different redshifts and distances and how those can be put together to derive empirically the evolution of galaxies and their structures. It finally concludes with a description of numerical computer simulations of galaxies that can be used to interpret the meaning of galaxy structure in terms of input physics.

To study the structural evolution of galaxies we are required to look back in time at the earliest galaxies and try to connect the systems we see to other galaxies through cosmic time. The distant universe itself is a vast topic with an already long history. Major advances in examining the structures of distant galaxies and how they connect to local galaxies is done primarily to date with the Hubble Space Telescope (HST), which was launched in 1990 (Figure 4.1).

Distant galaxies were known before the HST, although they were for the most part unresolved "blobs" of light. The history of distant galaxy research is such that astronomers are continuously surprised at how relatively easy it is to detect distant systems. It is not easy to do this, mind you, but compared to many who thought it impossible it is relatively straightforward using the largest telescopes.

The beginning of distant galaxy research really began with the discovery of quasars in the 1960s. Quasars were first discovered in the radio at Cambridge University and were later followed up with optical telescopes. These systems were bright in the radio, but simply appeared to be bright unresolved sources in the optical, hence the name quasi-stellar object (quasars) was adopted. At first, no one knew if these were in our galaxy or were outside of it. Likely no one thought that they were as distant as they eventually turned out to be.

doi:10.1088/2514-3433/abb602ch4

Figure 4.1. Image of the Hubble Space Telescope taken as it appears above the Earth. The HST has been the workhorse for discovering how galaxy structural evolution has occurred and will only be superseded by the James Webb Space Telescope (JWST). In many ways this telescope is the star character of this book. Credit: NASA/ESA. CC BY 4.0. (https://www.spacetelescope.org/images/hubble_earth/)

The key to figuring out of the nature of quasars was through their spectra. A major stumbling block to recognizing that quasars were actually distant systems was the fact that the spectrum had to be seen as one which is redshifted by such a large amount that that it took some time for it to be correctly interpreted. After this was shown to be the case by Schmidt (1963), the study of distant galaxies took off, and for decades quasars were the most distant objects in the universe.

However, it was not until the 1980s when studies of distant galaxies really started. This occurred when these system's redshifts were finally measured for example by Spinrad et al. (1985) who was, for example, continuously finding more distant galaxies. However this was done in a piecemeal fashion, and most of the distant galaxies were only one-off systems, and still only reaching around redshift of $z \sim 2$ by 1985. However, very few galaxies had such distances measured at those redshifts until well in the 1990s.

At the same time, but completely independently, large area imaging surveys of galaxies were finding a large population of blue galaxies that were very faint. The origin of these faint blue galaxies was unknown at the time, and this was considered a major problem in cosmology and galaxy formation (Ellis 1997). While individual bright objects could have their redshifts measured, the more common but fainter bluer galaxies were still a mystery. The idea was that these systems could be low mass bursting, but nearby dwarf systems, or they could be higher redshift but massive, forming systems.

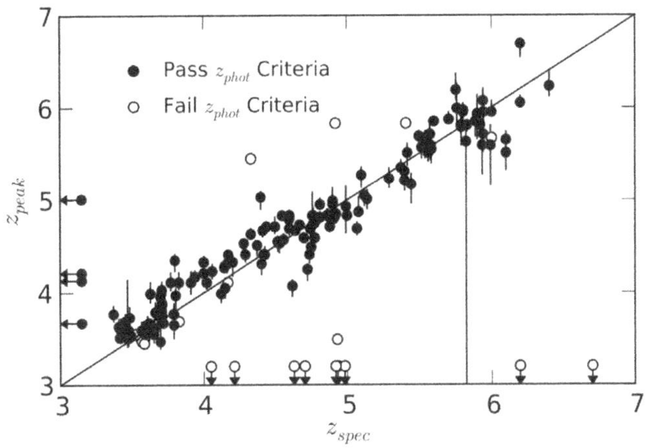

Figure 4.2. Plot of photometric redshifts versus spectroscopic redshifts. The quality of photometric redshifts is determined by how good this correlation is. These photometric redshifts are for those galaxies in the GOODS-N and GOODS-S fields taken from Duncan et al. (2014).

There were two new methods that came along in the early 1990s that changed this field dramatically. The first was the realization that because of the effects of intergalactic hydrogen gas the ultraviolet light from distant galaxies completely vanishes. From this, it is possible to locate distant galaxies by the fact that their UV light is extinguished by this gas that exists between the distant galaxy and us. Because these galaxies are discovered by this missing light in the UV, which is below the Lyman limit, these systems are now known as Lyman-break galaxies (LBGs). These systems were discovered by taking deep observations in the optical bands including the U-band where galaxies at redshifts above $z = 3$ would not have any discernible flux (Steidel et al. 1995).

The other big advancement was the development of so-called photometric redshifts, or photo-zs (Figure 4.2). We describe below in the Hubble Deep Field section how these are calculated. Because these photometric redshifts are based on photometry, or imaging, it is possible to obtain them for many galaxies in a relatively quick amount of time. This is why high fidelity photometry from the HDF and other Hubble fields really revolutionized this area. Photometric redshifts became a mature technique at a similar time as the LBGs in the mid-1990s. The method certainly came into its own after the HDF observations were released. One reason for this is that the HDF observations were based on four bands that covered the optical and UV light spectrum, and thus covered most of the spectral energy distributions for all the galaxies imaged (Figure 4.3). The success of these photometric redshifts with deep Hubble imaging made this a standard technique in distant galaxy studies.

The HST began taking pictures of the distant universe soon after it was launched, and this expanded greatly after the servicing mission in 1993 that effectively corrected the misfigured primary mirror due to spherical aberration, and the insertion of a new instrument COSPAR to fix this problem. After this occurred in

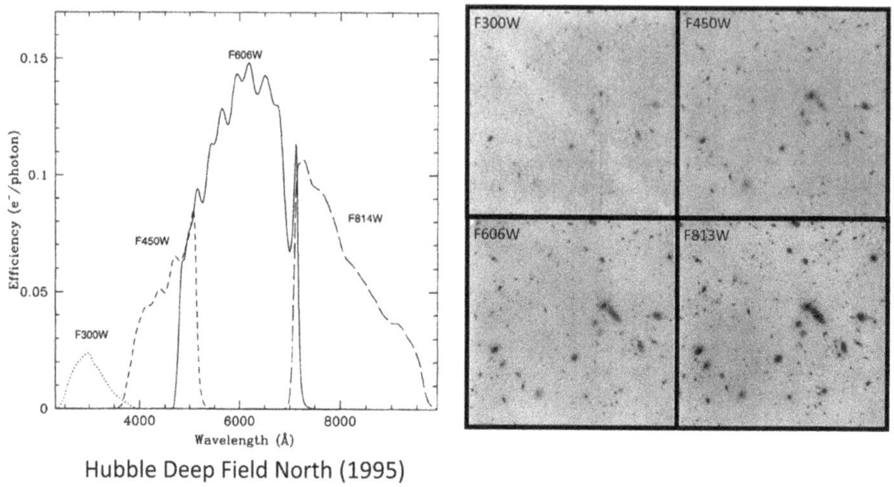

Figure 4.3. Parameters of the Hubble Deep Field. the left-side panel shows the sensitivity of the four HDF filters. The right picture shows one of the WF chips (WF3) in the four HDF filters: F300W, F450W, F606W, F814W. These images are compiled from Williams et al. (1996).

the 1993 servicing mission, the optics on Hubble were vastly improved giving very sharp images, meeting the initial specifications for the mission.

However, before the telescope was launched a question was asked about what Hubble would actually see once it took deep images of the universe. A paper by Bahcall et al. (1990) explored this, and other topics, and argued that it would not see very much in terms of the first galaxies. That is mostly due to theoretical ideas behind galaxy formation such that galaxies were thought to form very late in the universe. Therefore galaxies that existed early in the universe would be too faint to detect even with modestly deep exposures. This is due to the fact that within this idea these early galaxies would be very low-mass and thus very faint. There was in fact no plan to take ultra-deep or even modestly deep pictures of the same part of the universe once the HST was launched. It did not help that the theory said it would be more or less a waste of resources to attempt such things.

4.1 Pre-Hubble Deep Field Imaging with Hubble

How did the HDF idea to take deep pictures of a single pointing with Hubble develop then? The history of this is extremely interesting, and was developed due to a series of events that convinced the Director of the Space Telescope Science Institute (STScI) to undertake the HDF observations as part of his director's discretionary time. The history of this is nicely explained in Robert Williams' book on the HDF (Williams 2018).

When Hubble was first launched there were plans to carry out three key projects. One of these was a survey of blank fields of the sky, called the Medium Deep Survey (MDS). In many ways, the MDS was the precursor and justification for going even deeper with the HDF, and really showed what was possible to do with Hubble that could not be done from the ground—namely resolve the structures of distant galaxies and obtain high fidelity photometry for distant systems.

The strategy for the original MDS was to use the imaging camera on Hubble, called the Wide-Field Planetary Camera (WFPC), in parallel mode. That is, when one of the other instruments on Hubble took data on a target the WFPC camera would examine a blank field on the sky and expose for as long as the primary target was being observed with the other instrument. This is now a very common method for obtaining data with Hubble and random fields observed with HST are still used to find distant galaxies, sometimes even at very extreme distances. Even with the spherical aberration of Hubble, before the 1993 servicing mission, it was obvious that these images were providing new structural information for distant galaxies.

These early pre-refurbished HST images already showed irregularities even up to modest redshifts of $z \sim 0.4$ (Windhorst et al. 1994). These images had an effective resolution of 0.2″ but still could resolve these structures which were effectively a hint of things to come when the optics improved (e.g., Griffiths et al. 1994). It was also the case that the fainter galaxies in these fields were more likely to be peculiar and irregular, suggesting that there was indeed a redshift-morphology relation (e.g., Driver et al. 1995). However, at this stage in the mid-1990s there was little more than visual classification attempts to characterize the structures of irregular and peculiars and what these structures may mean in a quantitative and physical way for galaxy formation and evolution.

Other early observations included cluster fields such as those by Dressler et al. (1994) who by imaging the cluster CL 0939+4713 showed that distant galaxies imaged with Hubble were quite different from the systems we find in the local universe. At the time the redshift of this cluster at $z \sim 0.4$ was one of the most distant known. The galaxies that were seen with Hubble in these early days were quite clearly morphologically peculiar and irregular which was noted on nearly all early deep HST imaging starting from its first publications on this topic (e.g., Dressler et al. 1994; van den Bergh et al. 1996).

Dressler et al. (1994) was also one of the first attempts to explain these structures in terms of galaxy mergers and interactions in the context of the cluster environment. An even higher redshift cluster, surrounding the radio galaxy 3C 324, was imaged by Dickinson et al. (1995) using 34 HST orbits, the largest allocation to date for this type of deep imaging. It was clear that comparing Dickinson et al. and Dressler et al.'s imaging that the deeper one probed with Hubble, the imaging revealed fainter and more unusual galaxies. In general with deeper exposures more peculiar features could be seen within distant galaxies, revealing a new never before seen parameter space for distant galaxies. Hubble was demonstrating that for galaxy imaging, the deeper the better, and that even deeper imaging would hold even more surprises. This trend of obtaining clearly more peculiar galaxies at high redshifts has continued with deeper and redder HST observations, including those that sample the rest-frame optical in the observed near-infrared which are able to explore much higher redshifts than these early Hubble observations (e.g., Mortlock et al. 2013; Lee et al. 2013).

The next major advancement in looking at the structural evolution of galaxies began in very late 1995 with the HDF program. This imaging was a major advancement for the studies of galaxies and there has been nothing equivalent to its impact before or since. Essentially, the HDF was a project whereby the HST was simply pointed at a single part of the sky with WFPC2 and imaged this single area very deeply. The HDF was such an important observation in the history of this field it is worth going into it in more detail. In many ways, without the HDF this book would be written in an extremely different way.

4.2 The Hubble Deep Field

High redshift studies, especially for galaxy structure and evolution changed dramatically during the years 1995–1997 with the advent of both the HDF (Williams et al. 1996; Ferguson et al. 2000) and the discovery of a significant population of high-redshift galaxies. As recounted in Williams (2018) the inspiration for the Director to use his time to take these observations were the impressive imaging, already recounted in the previous section, taken of CL 0939+4713 and 3C 324.

The HDF itself was taken over 12 days from 1995 December 18 to December 30. The reason this program was so successful was that it carried out very deep observations in four filters: F300W, F450W, F606W, and F814W (Figure 4.3). In some ways these filters are the canonical filters used in the early days of HST for galaxy studies, and the HDF help made them so. The depths, exposure times, and 10 σ depths for these filters are: F300W for 153700s, reaching 26.98AB; F450W for 120600s, reaching 27.86AB; F606W for 109050s, reaching 28.21AB; and F814W for 123600s, reaching 27.60AB. The deepest bands are in the redder part of the spectrum, with the deepest imaging within the F606W band (Williams et al. 1996). A combined color version of the HDF is shown in Figure 4.4.

The long wavelengths probed in the HDF imaging effectively meant that distant galaxies could be discovered by two related techniques. The first is the Lyman-break technique (Steidel et al. 1995; Shapley 2011) which finds galaxies that are often referred to as LBGs, as described earlier. LBGs are fairly simple systems in that they are identifiable by the fact that their UV light below the Lyman-break is absorbed by intergalactic hydrogen and hydrogen in the outer parts of the galaxies themselves. This produces an identifiable "break" where the light from the galaxy appears to disappear in some filter, only to reappear in a slightly redder filter. For traditional LBGs, this differential is large—that is above the break the galaxy is very bright, so effectively as we go from blue to red wavelength the image of the galaxy goes from almost nothing to a very bright system which cannot be produced by known stellar populations—thus the conclusion is that these systems have their UV light extinguished by intervening hydrogen gas. This idea is verified by examining the actual redshifts of these systems through emission lines, which backs up the redshifts derived through the Lyman-break.

The other method for determining the distances to these galaxies is to use the already introduced photometric redshifts, of which the Lyman-break technique is a

Figure 4.4. The original Hubble Deep Field which was taken in late 1995 and released to the community in 1996. Very likely one of the most important images in astronomical history. Within this picture are around 1000 galaxies at various distances from relatively nearby (including a few stars) to those that formed when the universe was only about 1 billion years old. Within this image shows about 13 billion years of cosmic evolution with, in general, the fainter and bluer objects the more distant galaxies. Credit: Robert Williams and the Hubble Deep Field Team (STScI) and NASA/ESA. CC BY 4.0. (https://www.spacetelescope.org/images/opo9601c1/)

simplified version of. Photometric redshifts (often called photo-zs) are calculated by using the photometry of the galaxy and fitting this to model spectral energy distributions for galaxies of different types and at different redshifts. These photo-zs are based on the broad-band photometry of galaxies. The basic idea is that model spectra of galaxies of various star formation histories or galaxy types are convolved with the filter responses for the broad-band photometry obtained. Essentially either a machine learning technique, or a best-fit method is employed to obtain the photometric redshift by comparing the model spectral energy distributions with the actual photometry. The quality of these photo-zs are then obtained by comparing the photometric and spectroscopic redshifts (Figure 4.2).

In the simplest methods, there is a two-dimensional space of redshift and galaxy spectral type that ranges from galaxies that contain all old stars formed early in the universe to galaxies whose light is dominated by recent star formation in the last 100 Myr before it was observed. Within this two-dimensional parameter space most galaxies can be uniquely identified, thereby providing the galaxy type as well as its redshift through the fit. Ultimately, these photometric redshifts have to be calibrated by spectroscopic redshifts which are measured typically from emission lines. The quality of these photo-zs can be very accurate (e.g., Dahlen et al. 2013).

The HDF, and later campaigns to acquire very deep Hubble imaging in small fields includes in rough order in which they were carried out: the Hubble Deep Field South (Williams et al. 2000), the Great Observatories Origins Deep Survey (GOODS; Giavalisco et al. 2004), the Hubble Ultra Deep Field (UDF; Beckwith et al. 2006), the COSMOS field (Scoville et al. 2007), the Extended Groth Strip survey (EGS; Davis et al. 2007), and most recently the CANDELS survey (Grogin et al. 2011; Koekemoer et al. 2011) and the Hubble Frontier Fields (Lotz et al. 2017). These surveys have all in many different ways revolutionized the field of galaxy formation studies and in particular the study of galaxy structure and morphologies which we describe in detail in the later parts of this book.

The critical aspect within these deep fields is not that they are simply deeper than previous deep HST imaging, but that they are the focus of large-scale efforts to obtain photometry at the faintest levels at nearly all wavelengths. First of all, this allowed redshifts to be measured for most galaxies using photometry through the previously described photometric redshifts. These redshifts allow us to measure evolution over large swaths of time which was not possible with spectroscopic redshifts which were, and still are, hard to obtain for many systems. Spectroscopic redshifts can be obtained in bulk with instruments such as large multi-slit spectrographs such as LRIS on Keck (e.g., Cohen et al. 1996) or the VIMOS spectrograph on the VLT (Le Fèvre et al. 2005). However, even when there are an abundance of spectroscopic redshifts, the samples obtained from them are often biased due to the brightest and most star-forming galaxies which are the easiest to obtain spectroscopic redshifts for.

Thus, early HST observations including an initial look at the HDF showed that fainter galaxies were irregular/peculiar, but there was still no known information on how far away these systems were, let alone their nature. Luckily, early observations using a mixture of spectroscopic and photometric redshifts revealed that many of these peculiar galaxies were quite distant and that the Hubble sequence was certainly not in place at high redshifts $z > 1$ (e.g., van den Bergh 1996; Figures 4.5 and 4.6). One issue with these early observations was that by using WFPC2 the reddest filter was at wavelengths around 9000Å, which means that the rest-frame light probed by all filters at higher redshifts than about $z \sim 1$ was in the ultraviolet. However, it was clear from these observations that galaxies were irregular and peculiar and fainter galaxies more so than brighter (e.g., Driver et al. 1995).

The fact that these galaxies are intrinsically peculiar and irregular was verified in the rest-frame optical after deep near-infrared observations of the HDF were taken

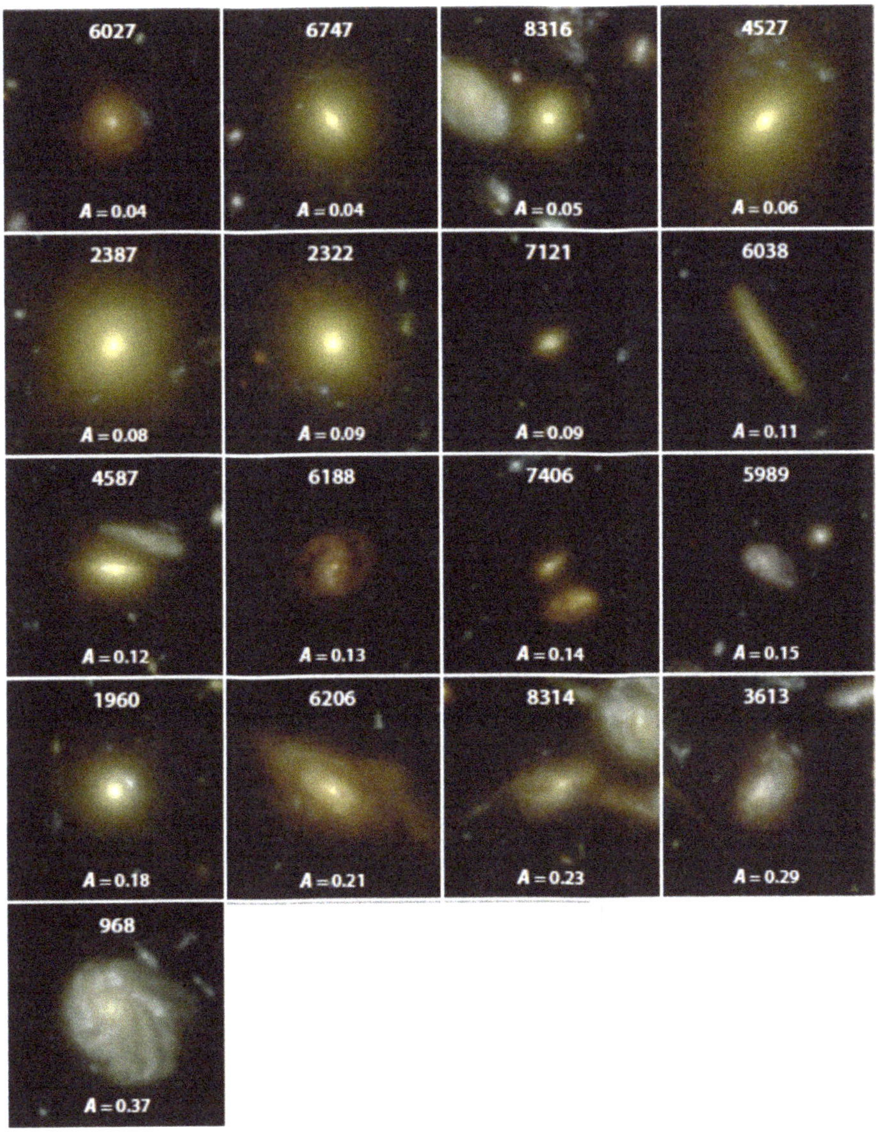

Figure 4.5. Lower redshift galaxies found within the Hubble Ultra Deep Field (UDF) as taken by the ACS camera, ordered by their asymmetries. These systems are selected with redshifts $0.5 < z < 1.2$ and with stellar masses $M_* > 10^{10}\ M_\odot$. The ID is the number used in Conselice et al. (2008), and the A value is the value of the asymmetry. At these redshifts most of the massive galaxies can still be classified as being on the Hubble sequence.

in 1998 with NICMOS (Dickinson 2000). An analysis of this image showed that using the rest-frame optical to study the structures of galaxies revealed that the Hubble sequence was nearly completely absent at $z > 2$—that is, very few galaxies at these redshifts could be classified as an elliptical or spiral. Through detailed observations it was shown that only at $z \sim 1.5$ did spirals and ellipticals start to

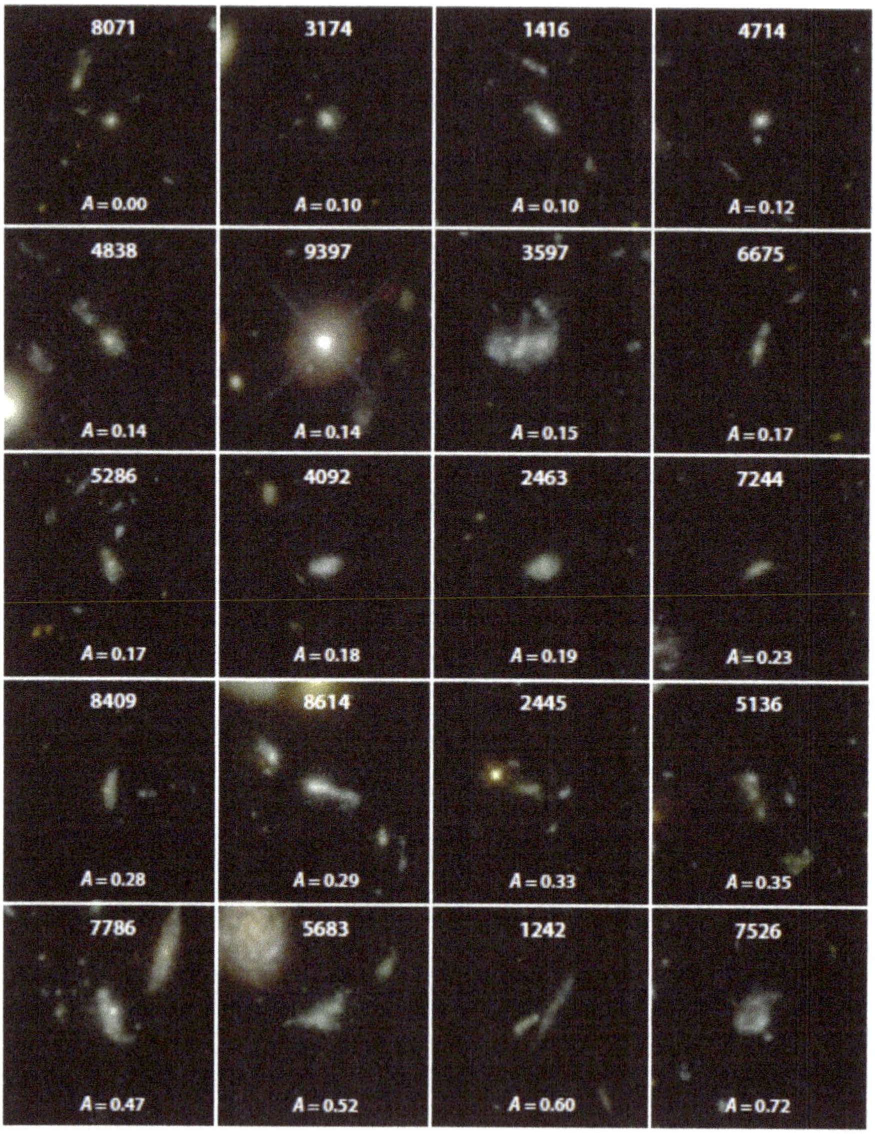

Figure 4.6. Images of massive galaxies in the Hubble Ultra Deep Field (UDF) imaged by the ACS camera, ordered by the value of their asymmetries from the most symmetric to asymmetric. The galaxies shown here are systems with stellar masses $M_* > 10^{10}\ M_\odot$ at redshifts $2.2 < z < 3$. Galaxies at these higher redshifts are typically much smaller, bluer, and have more distorted morphologies and higher measured merger fractions than galaxies of comparable mass today (Conselice et al. 2008).

become as common as the peculiar and irregular galaxies (Conselice et al. 2005; Buitrago et al. 2013; see also Chapter 5).

4.3 Galaxy Visual Morphology Evolution

Some of the most up to date results of this type of examination on the evolution of visual morphology, as measured by visual types, originates from the CANDELS

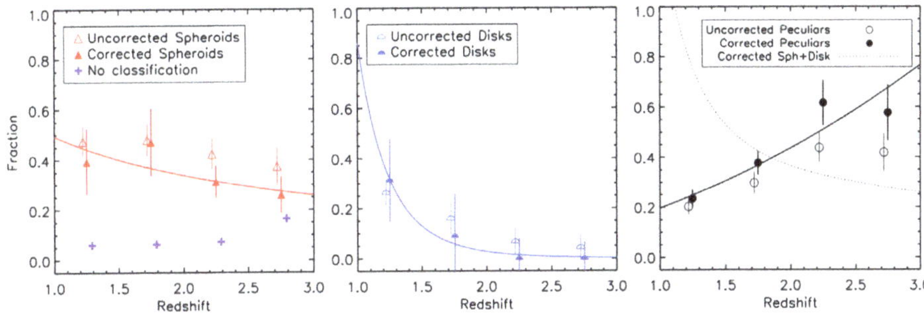

Figure 4.7. This plot shows the corrected evolution of the fraction of the different galaxy types with redshift. What is plotted is such that the empty points are the original fractions and the filled points are the fractions corrected as discussed in the text. The x-axis values are shifted by a small amount for clarity. From Mortlock et al. (2013) © 2013 The Authors. CC BY.

survey. One version of this is shown in Figure 4.7 (Mortlock et al. 2013) based on visual classifications for galaxies with stellar masses $M_* > 10^{10}\ M_\odot$. It is important to stress that this figure only reveals a visual estimate of galaxy morphology, and does not, in itself, reveal with definiteness what the actual 'type' the galaxy is. That is, the classification of a spiral, elliptical, or peculiar does not in itself mean that the galaxy classified as such has a star formation rate, color, mass, or size expected for similar morphologically classified galaxies in the nearby universe. Detailed examinations show that what is often the case is that these visual morphologies do not correlate well with other physical properties (e.g., Conselice et al. 2011; Mortlock et al. 2013). There is also a stellar mass differential in the way that the Hubble sequence forms. This is such that the highest mass galaxies appear to form into visual ellipticals and disks before the lower mass systems do (Mortlock et al. 2013). Figure 4.7 clearly shows that at higher redshifts, $z > 2$, galaxy types are dominated by peculiar galaxies. At lower redshifts disks and ellipticals are more common (Conselice et al. 2005; Delgado-Serrano et al. 2010; Mortlock et al. 2013; Buitrago et al. 2013). These morphologies have been corrected by image simulations, such that galaxies will have a different morphology if they are imaged at higher redshifts (Figure 4.8). Figures 4.5 and 4.6 are image montages from Hubble imaging of the UDF of the most massive galaxies at both $z < 1.2$ and $z > 2.2$. These images show the stark differences between the two types. We later examine in Chapter 5 what these peculiars have potentially evolved into at lower redshifts. There is however a real trend whereby peculiar systems have a higher star formation rate compared with smoother galaxies. This trend continues and is seen for galaxies with internal features such as blobs (e.g., Lee et al. 2013). All of this information reveals that while the Hubble sequence itself was not formed early in the history of the universe, there are trends which suggest it became established when the universe was half of its current age.

These plots are simply demonstrating the fractional evolution of galaxy types with redshift. What we would also like to know is the number density evolution for galaxies with different morphologies and types. From mass measurements we now know that the most massive galaxies are for the most part in place at $z \sim 1$ with a similar number density as at $z \sim 1$ (e.g., Bundy et al. 2005; Mortlock et al. 2015).

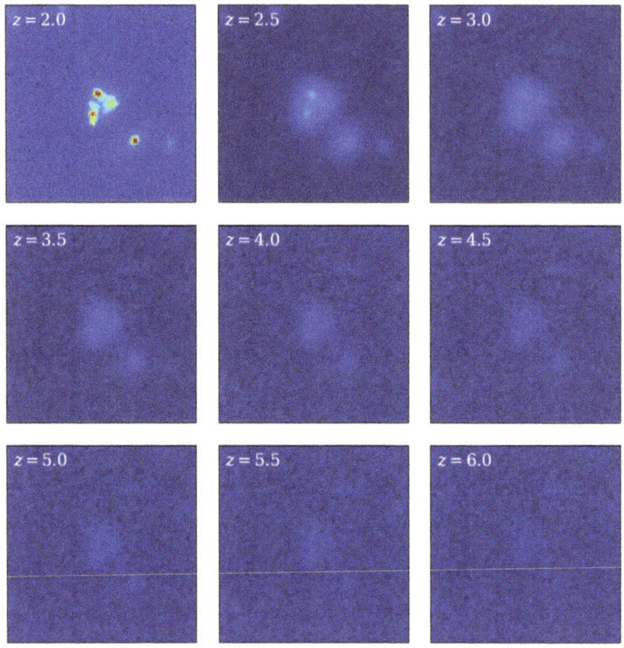

Figure 4.8. A galaxy simulation to very high redshifts showing how even a bright galaxy at $z \sim 2.0$ would become progressively more difficult to detect and study at higher redshifts. Reproduced from Whitney et al. (2020).

At the same time, it is the lower mass galaxies that typically increase the most in terms of number density and fraction over the redshift range $z \sim 0$–2. In the local universe most massive galaxies are elliptical in morphology, implying that most of the evolution in morphology occurs later in the history of the universe for the lower mass galaxies, which are more irregular/peculiar than the higher mass galaxies at $z < 1$. Thus, in addition to a mass downsizing, there exists a morphological downsizing, such that massive ellipticals are formed before other galaxy types, most notably spirals.

4.3.1 The Formation of Galaxy Structure

One of the biggest questions with galaxy morphology is answering when modern spirals and ellipticals form. Of course, this is directly related to the formation of the Hubble sequence. However, this is usually not so simple, as what appears as an elliptical/spiral may be quite different from systems classified as such in the local universe. This is not so much in terms of morphology, but in terms of physical properties such as ongoing star formation rate, color, or stellar populations. To really answer when galaxies of a certain type form we must determine when systems with the same morphology as local galaxies (i.e., disks and ellipticals) achieve a similar physical state determined through other properties in addition to morphology, such as color or star formation rate. With these criteria, we find that very few analogs of modern ellipticals and spirals exist at higher redshifts.

Morphological fractions of different galaxy types are similar at $z \sim 0.5–0.8$ compared to the fractions of these types in the nearby universe. However, when we examine the physical properties of galaxies within the same type (for example within those classified as ellipticals)—we see a diversity in physical properties such as colors, star-forming knots, and tidal features. These properties have not evolved yet to reach the same level or values for local galaxies of the same morphological type. It is only relatively recently at $z < 0.3$ where we see that the galaxy population appears similar to that of today in most properties, even beyond morphology. This allows us to conclude that galaxy morphology and structure are long-lived features in galaxies and are not transient. They become established before the stellar populations of systems in the distant universe match what would be expected for their morphological types corresponding at $z \sim 0$ (Conselice et al. 2011).

The question of how morphology is established can be subdivided into many questions. Some of these include how and when features such as bars and rings, and spiral patterns assemble. These are the topics which we discuss in the next few sections.

4.4 Galaxy Bar Evolution

A primary consideration when examining the evolution of galaxy structure is not just the gross morphology, but also when sub-features which are important, such as bars, form. The evolution of the bar fraction with time is one of the most interesting of these as bars are a fundamental part of the Hubble sequence and are thought to be the way in which star formation and AGN are possibly triggered. Historically, early studies found that the bar fraction changes significantly with time (Abraham et al. 1999), while others have found that bars were already present in a similar fraction as today at $z \sim 1$ (Elmegreen et al. 2004; Jogee et al. 2004). On the other hand, Sheth et al. (2008), using the HST COSMOS survey, found that the bar fraction increases over $z = 0.84$ to $z = 0.2$ from 20% to ~60% of disk galaxies. There is however a mass dependence on this evolution, whereby Sheth et al. (2008) find that the bar fraction is within the errors constant with redshift for the most massive and red disk galaxies, but evolves more for the lower mass systems. Most of the observed evolution in bars occurs within the population of bluer lower mass disk galaxies. This is another sign of galaxy downsizing, whereby the most massive systems appear to take on their local equivalence before lower mass galaxies. This is thus seen often in the morphology and structure as well.

The problem with these previous studies is that they are based on observed optical imaging, typically with the ACS camera on the HST or with very small field of view imaging with NICMOS. With the advent of the WFC3, large scale imaging in the near-infrared imaging was possible, particularly within the CANDELS fields. As part of this Simmons et al. (2014) identified 123 barred galaxies, finding a fraction of ~10% over all redshifts, with no significant evolution up to $z \sim 1.5$. Thus, it appears that disk galaxies have had bars in a similar proportion for about 8 Gyr in the same fraction as disk galaxies do today (Figure 4.9).

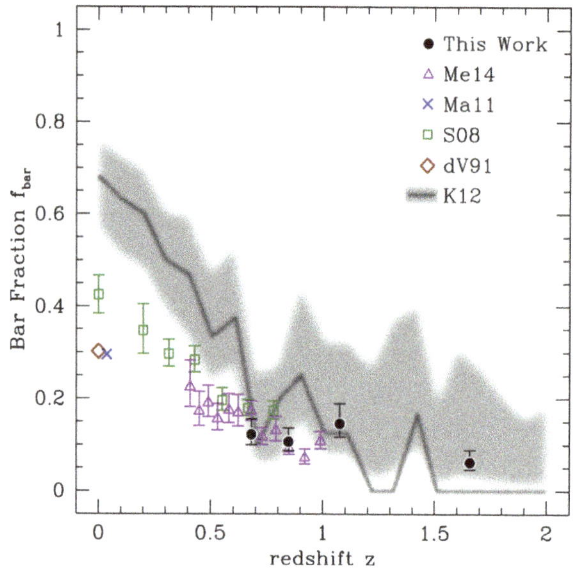

Figure 4.9. Plot showing the changes in the bar fraction for spiral galaxies a function of redshift, taken from Simmons et al. (2014). Also shown are the results from Sheth et al. (2008, S08, green squares), Melvin et al. (2014, Me14, purple triangles), Masters et al. (2011, Ma11, blue cross), and de Vaucouleurs et al. (1991, dV91, red diamond).

Bars forming over time in spiral galaxies reveal when disks become dynamically mature enough to form these structures. The presence of bars also allows us to make inferences about the role bars play in the evolution of bulge formation, star formation, and the growth of central massive black holes through the triggering of AGN. The bar studies with time mentioned above all locate bars within disks based on the measured changes in ellipticity and position angle in the surface brightness profiles of galaxies, or they are based on visual inspections.

4.5 The Formation of Spiral Structure

One of the most interesting features of deep imaging is that we can probe the formation of the first throws of formation for the Hubble sequence—namely the formation of spiral galaxies and ellipticals. This is more easily done for spirals in some ways as these are easier, but not trivially so, to distinguish between elliptical galaxies. It is important to note that spiral galaxies and disk galaxies differ, and are not interchangeable. Disk galaxies are those that are supported by rotation and have a flattened structure. Spiral galaxies are those disks that have spiral structures, such as those shown in Figure 1.4.

An issue related to bars is the onset of spiral structure, which has remained a problem for decades. The formation of spiral structures relates to fundamental physics for explaining how they form, evolve, and remain over cosmic time. Spiral structures are also important for a variety of other reasons, such as triggering of star formation episodes and possibly interactions with other galaxies.

The best location to carry out this type of analysis to date is within the deepest HST imaging, namely the large area CANDELS fields that have been described already. Within these fields, we can find candidate spiral galaxies through the determination of morphology types based on visual estimates. This may not be the best way to carry out this type of analysis, as visually galaxies can look very unusual at high redshifts, but it is a good first way to address this question as even a rotating system is not necessarily a disk or spiral galaxy. Also, spiral galaxies are symmetrical typically and traditionally have been identified by eye.

Various numbers of spiral galaxies have been confirmed to exist at high redshifts based on the appearance of spiral structure (e.g., Bruce et al. 2012; Margalef-Bentabol et al. 2018; Figure 4.10) as well as some systems containing evidence for rotation (e.g., Law et al. 2012). It is however especially difficult to be sure of an unambiguous detection of spiral structure in distant galaxies due to resolution/depth issues.

Several papers based on ACS (pre-WFC3) data from HST showed that there are many examples of disk-like morphological systems at high redshifts. This includes early work by Labbè et al. (2003), the Luminous Diffuse Objects of Conselice et al. (2004), and systems identified by Elmegreen et al. (2007) as clump-clusters. Law et al. (2012) discovered a galaxy with a likely spiral structure at $z = 2.18$, which also hosts a large internal velocity dispersion from IFU spectroscopy. Spiral arms are for the most part very rare at high redshift. It is likely that only a small number are formed in disk galaxies before $z \sim 1$.

Another feature of spirals and disks is the formation of bulges. The traditional scenario is that bulges form in mergers of two galaxies of similar mass. However,

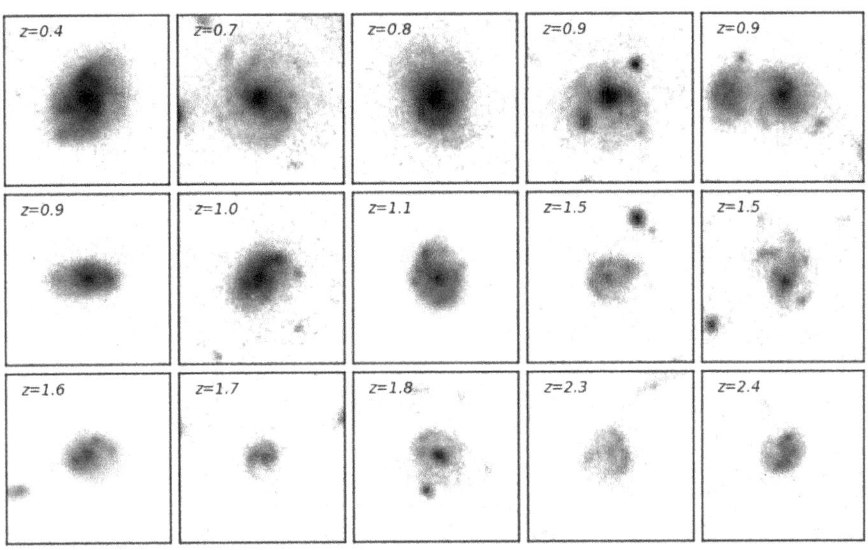

Figure 4.10. Examples of high-z spiral galaxies as found in the CANDELS fields in Margalef-Bentabol et al. (2020, in preparation). These are disks and spiral galaxies in formation seen at the highest redshifts where these kinds of galaxies can be accurately probed currently.

recent results suggest that bulges can and probably do form in multiple ways. The so-called pseudo-bulges differ from classical bulges as they are likely formed through secular processes, such as through bars, within a disk. Some proof of this comes from the correlations found between bulge properties and central massive black holes.

Due to the higher resolution NIR imaging from CANDELS it has been possible to examine this problem and determine how bulge to disk ratios have evolved for galaxies as a function of redshift (e.g., Bruce et al. 2012; Margalef-Bentabol et al. 2016, 2018). What is found within these studies is that there is an evolution in the bulge to disk ratio such that it increases with lower redshifts, and again seems to do so earlier for high mass galaxies than for lower mass ones (Figure 4.11). There is also a corresponding growth in the disks as opposed to the bulges (e.g., Margalef-Bentabol et al. 2018).

As a galaxy's apparent morphology can often be ambiguous in terms of matching with a Hubble type, spectroscopy can often help us address this issue. A classical idea is that ellipticals are supported by isotropic motions with no large net motion, while galaxies which are disk dominated are thought to be rotationally dominated. While this type of separation is no doubt too simplistic, it can give us some guidance to the observations. Integral field unit (IFU) or long-slit spectroscopy can help us understand, study, and locate the presence of bulges and disks. Most recently, integral field spectroscopy on $z > 1$ systems reveals important clues about the nature of high redshift galaxies, in particular peculiar ones (Förster Schreiber et al. 2009; Glazebrook 2013; Buitrago et al. 2014; Harrison et al. 2017). Some of the biggest studies have been carried out with the SINFONI IFU and KMOS IFU on the VLT, as well as work done with OSIRIS and MOSFIRE on Keck (Law et al. 2012; Price et al. 2020).

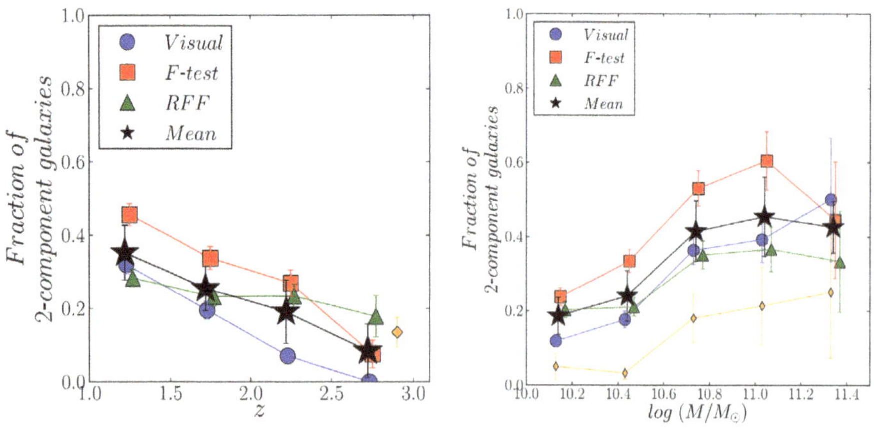

Figure 4.11. Figures from Margalef-Bentabol et al. (2016; © 2016 The Authors. CC BY) regarding the growth of bulges and disks as a function of redshift up to $z = 3$. The left shows the fraction of galaxies which are in two-component systems, as opposed to single systems; this is determined through various tests that show a galaxy is better fit by two components than by a single component, or vice-versa. The right-hand panel shows the fraction of galaxies which are better fit by these two components than one—showing that higher mass galaxies are more likely to need two components, demonstrating that these massive systems are undergoing more structural evolution than lower mass systems.

These surveys including, e.g., Förster Schreiber et al. (2009) find a roughly equal distribution of different kinematic classes which are: rotationally dominated, mergers, and compact galaxies with large velocity dispersions (e.g., Buitrago et al. 2014). The real nature of these systems, and their formation, is not yet known for some, although clumpy galaxies at $z > 2$ tend to have a high velocity dispersion, which is a sign that they originate from intense star formation or merging activity. In the future, progress can be made to infer the origin of these structures and kinematics by combining larger surveys of IFU measurements with imaging to decipher physical states and origins.

However, most of what we have discussed in this section relates to what appears to be going on with galaxies in terms of their apparent structure measured by eye. In the next chapter we will be more quantitative and measure evolution directly with measurements of quantified structural parameters.

4.6 Numerical Simulations of Galaxy Structure

Simulations of galaxy formation are extremely useful and often needed to understand the meaning of galaxy structures and how to interpret them. This is especially the case for understanding how to use structural indices which are explained in more detail in the previous chapter for measurement and the next chapter for use in terms of measuring evolution.

Work on this began relatively early, long before any deep observations with Hubble were taken. Perhaps the most famous early work was the computer simulations by Toomre & Toomre (1972) who demonstrated that the peculiar morphologies of galaxies that were seen in observations such as the, e.g., the Arp (1966) atlas were due to systems undergoing major mergers rather than some other cause. These observations of peculiar galaxies produced by mergers in a simulation was a major result at the time, as it was unknown what process could produce the disturbed structures seen in some galaxies. Since that time numerical simulations of galaxies have become an essential method for understanding the structures and morphologies of galaxies in both the local and distant universe. This is also a fast developing field, with simulations becoming rapidly more advanced and sophisticated over time.

These types of simulations continued to improve, and became very sophisticated for simulating galaxy mergers in the 1990s. Much of this work was influenced by the peculiar and irregular morphologies seen in deep HST imaging that was being acquired at the same time. For example, Mihos & Hernquist (1996) demonstrates in great detail how peculiar galaxies seen in deep HST pictures could be due to the merger process using the TREESPH hydrodynamical method from Hernquist & Katz (1989). Much of the interest in these simulations was based on comparing simulated merger results with detailed observations. This was carried out with simulations using starbursts spectral energy distributions (e.g., Barnes & Hernquist 1991). At the same time these N-body simulations of mergers were used as a method of interpreting HST morphologies of galaxies as mergers (Mihos 1995).

4.6.1 Merger Timescales with Structure

The use of quantitative structural methods described in Chapter 3 applied to these N-body and hydrodynamical simulations of structural evolution was first carried out by Conselice (2006) to measure merger timescales. Conselice (2006) used the CAS method to calculate the timescale of the merger process on these dark-matter particle simulations. This timescale in which CAS and other indices are sensitive to is critical for interpreting galaxy merger fractions (described in Chapter 5) through cosmic time. These timescales allow us to convert merger fraction to merger rates, and from that calculate how mergers are driving galaxy formation. These early simulations did not include star formation, only the dark matter, so there are some limitations to these results, but others followed up this work with more detailed simulations, finding similar results (e.g., Lotz et al. 2008). Using these dark matter simulations of both major and minor mergers, Conselice (2006) find a merger timescale of ~0.3–0.8 Gyr for the time period of galaxies undergoing mergers to be identified as such through the CAS criteria (Chapter 3).

Conselice (2006) also demonstrated that the time when an ongoing merger is asymmetric enough to be identified through the CAS system is distributed throughout the merger process, and is not located at one particular time (Figure 4.12). They also find that merging simulated galaxies would not always be identifiable as mergers when examined within CAS parameters. Conselice (2006) calculate that only during a fraction (around 1/3) of a merging galaxy's time would it be found in the merger region of CAS space. This simulation and others showed that the timescale is largely independent of the viewing angle of the merger. That is, viewing a galaxy at different orientations does not have a strong effect on the resulting measured merger timescale, which is what one would want for a robust identifier. Conselice (2006), and others such as Lotz et al. (2008), found that the asymmetry signal is also only present during major mergers when the mass ratios are 1:3, or greater. That is, minor mergers with ratios of less than 1:3 do not produce a high asymmetry signal that would be identifiable as a merger.

The simulations of Conselice (2006) are however simplistic as they do not include the effects of star formation or dust, which can produce dramatic changes in morphology (e.g., Taylor-Mager et al. 2007). When star formation and dust are added to galaxy merger simulations the quantitative structural parameters measured are similar to those seen in nearby galaxies. It is also the case that these measured parameters correlate with other properties such as color in a similar way as it does for nearby galaxies (e.g., Hambleton et al. 2011).

Lotz et al. (2008, 2010a, 2010b) measured CAS and Gini/M_{20} parameters on numerical simulations including old and young stars, star formation, gas, and dust. These simulations use GADGET/N-body/hydrodynamical simulations, which are then used to image the evolving structures of galaxies in various mergers. Lotz et al. (2008) examined this for disk galaxies of similar total mass, while Lotz et al. (2010a) investigated mergers with a range of mass ratios. Furthermore Lotz et al. (2010b) examined how observables change with different amounts of cold gas in the merging galaxies and how that influences the structure and morphology. These simulation

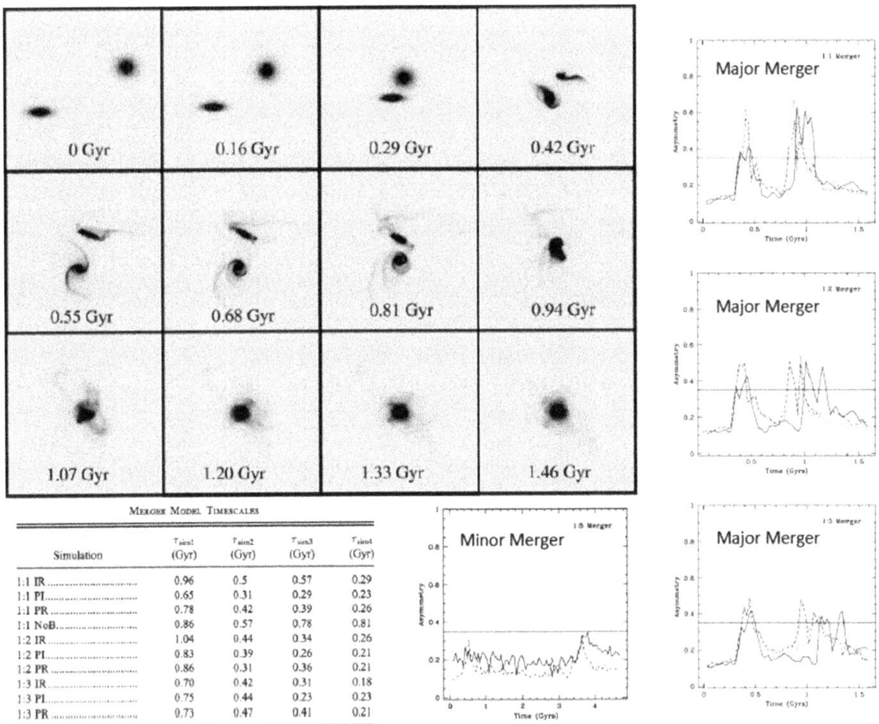

MERGER MODEL TIMESCALES

Simulation	T_{sim1} (Gyr)	T_{sim2} (Gyr)	T_{sim3} (Gyr)	T_{sim4} (Gyr)
1:1 IR	0.96	0.5	0.57	0.29
1:1 PI	0.65	0.31	0.29	0.23
1:1 PR	0.78	0.42	0.39	0.26
1:1 NoB	0.86	0.57	0.78	0.81
1:2 IR	1.04	0.44	0.34	0.26
1:2 PI	0.83	0.39	0.26	0.21
1:2 PR	0.86	0.31	0.36	0.21
1:3 IR	0.70	0.42	0.31	0.18
1:3 PI	0.75	0.44	0.23	0.23
1:3 PR	0.73	0.47	0.41	0.21

Figure 4.12. Figure showing the results of pure N-body model simulations taken from Conselice (2006), demonstrating the structure of two equal size disk galaxies merging as a function of time. The numbers on the top of the realizations of this model show the various snap shots of time through the simulation. The right panels show the changes in the asymmetry values as a function of time for different simulations, with 1:1 mass ratio mergers shown on the top right, and two simulations with 1:3 mass ratios below this. The middle bottom plot shows the asymmetry changes for a minor merger with a mass ratio of 1:10. The horizontal line in these plots shows the limit for finding mergers, with mergers those systems above the line. This demonstrates the changing form of quantitative indices during a merger, and how these systems are only identifiable within the different systems as a merger at specific times. Reproduced from Conselice (2006).

results are then passed through the SUNRISE Monte-Carlo radiative transfer code to produce, as realistically as possible, how galaxies would appear based on the simulation output.

Similar to Conselice (2006), Lotz et al. (2008, 2010a, 2010b) investigate the location of galaxies and how they evolve in CAS and Gini/M_{20} parameter space in different scenarios, and for different properties of the merging galaxies. They investigate the timescale for how long simulated galaxies appear as a "merger" based on where they fall in these non-parametric structural spaces (see Figures 3.1 and 3.3). In these papers the effects of dust, viewing angle, orbital parameters, gas properties, supernova feedback, and total mass are tested to see how they alter the measured merger timescale. It is found that most properties—the total mass, supernova feedback, viewing angle, and orbital properties of mergers have very little influence on the derived timescales. This is another good sign that these parameters are robust and stable to features that are secondary to the merger stage. However, what is found is that the mass ratio and gas mass fraction can have a significant influence on the derived merger timescales.

These simulations show that mergers are identified within both CAS and Gini/M_{20} space at the first pass of the merger when the two galaxies begin to merge. They become identifiable mergers again when the systems finally coalesce to form a remnant (Conselice 2006; Lotz et al. 2008). However, not all phases of simulated merging galaxies are located in the merger area of non-parametric structural parameter space. However, this methodology allows the timescales for mergers, as measured from structure, to be calculated. For example, Lotz et al. (2008, 2010a) find that the CAS asymmetry timescales for gas-rich major mergers are about 0.2–0.4 Gyr up to mass ratios of 1:3. For minor mergers, the timescale is very short using asymmetry at 0.01–0.06 Gyr (Lotz et al. 2010a). This means that a minor merger does not produce a high enough signal for long enough to be identifiable as a major merger, consistent with Figure 4.12.

The merger timescale for Gini/M_{20} merger identifications of major mergers ranges from $\tau_m \sim 0.2$–0.4 Gyr, similar to the CAS timescale. On the other hand, the timescale for minor mergers within the Gini/M_{20} system ranges from 0.2 to 0.5 Gyr, showing that Gini/M_{20} is more likely to find minor mergers and major mergers mixed together than the asymmetry index. This means that asymmetry is much better for measuring cleaner samples of major mergers, and Gini/M_{20} for all mergers. Lotz et al. (2010a) demonstrate this and conclude that the CAS methodology locates primarily major mergers, whereby the Gini/M_{20} method finds a mixture of minor and major mergers. In this sense, it is possible to use both methods to get a fuller idea of different types of ongoing mergers in a galaxy sample. These timescales can then be used to convert observed merger fractions into merger rates, as discussed in the next chapter.

One caveat is that Lotz et al. (2010b) show that very gas rich galaxies, such as likely ones at high redshift may have longer timescales for being identified as merging systems. If correct, these "merger" asymmetry signals for minor mergers with gas rich galaxies could in principle mimic what is seen for major mergers without as much gas. For this reason, care must be taken when applying the Gini/M_{20} index to high redshift galaxies. What we do know however is that that massive galaxies, even at very high redshift, with $M_0 > 10^{10}$ $M_\odot$, have a low gas mass fraction (e.g., Erb et al. 2006; Mannucci et al. 2010), thus it is unlikely that this will affect the inferred merger timescales for massive distant galaxies.

4.6.2 Dynamical Friction Calculations for Pairs of Galaxies

In addition to using structure, we can utilize calculated merger timescales from dynamical friction, which correspond to using galaxies in pairs, to determine the merger rate (see Chapter 5). Using these calculations we find that the calculated merger timescale within pairs under dynamical friction is used to determine how long it takes for two galaxies to change their separation from r_i to r_f. This dynamical friction timescale in an analytical formula is given by, $t_{\rm fric}$,

$$t_{\rm fric} = 0.001\ 4{\rm Gyr}\left(r_i^2 - r_f^2\right)\left(\frac{v_c}{100\ {\rm kms}^{-1}}\right)\left(\frac{10^{10}\ {\rm M}_\odot}{M}\right), \quad (4.1)$$

whereby v_c is the relative velocity between the two merging galaxies at a given time, M is the mean accreted mass, and the Coulomb logarithm, which for most mergers is $\ln \Lambda = 2$ (Dubinski et al. 1999). The very basic calculation given in Equation (4.1) is very often a reasonable estimate for merger timescales, giving values of ~0.5 Gyr for equal mass mergers. We compare the merger rates with structure and pairs using these timescales in the next chapter. We will also discuss the use of this dynamical friction calculation in more detail in that chapter as well.

More recently it has become clear that galaxies merging at higher redshifts have a shorter timescale based on calculations from simulations such as the Illustris simulation (e.g., Snyder et al. 2017; Conselice et al. 2020, in preparation). This is such that the timescales for galaxies merging declines with redshift as $\sim(1 + z)^{-2}$, meaning that the merger rates will be higher at higher redshifts than when using a simple constant merger timescale. Furthermore, empirical measurements of looking at the merger rate by examining the changes of the merger fraction with time at lower redshifts find a time-scale of 0.6 ± 0.3 Gyr at $z < 1$ (Conselice 2009). This is in good agreement with the numerical and dynamical friction calculations, and suggests that we are measuring as accurately as possible the merger timescales for galaxies at least up to $z \sim 3$.

These numerical and hydrodynamical simulations are useful in many instances besides merger-timescales. In later chapters we will further describe their ability for deriving physical properties of galaxies from galaxy structure changes with redshift. This is a very rapidly growing field of extragalactic research and comparisons to observations are really still just beginning.

References

Abraham, R. G., Merrifield, M. R., Ellis, R. S., Tanvir, N. R., & Brinchmann, J. 1999, MNRAS, 308, 569

Arp, H. 1966, ApJS, 14, 1

Bahcall, J. N., Guhathakurta, P., & Schneider, D. P. 1990, Science, 248, 178

Barnes, J. E., & Hernquist, L. E. 1991, ApJL, 370, L65

Beckwith, S. V. W., Stiavelli, M., Koekemoer, A. M., et al. 2006, AJ, 132, 1729

Bruce, V. A., Dunlop, J. S., Cirasuolo, M., et al. 2012, MNRAS, 427, 1666

Buitrago, F., Trujillo, I., Conselice, C. J., et al. 2013, MNRAS, 428, 1460

Buitrago, F., Conselice, C. J., Epinat, B., et al. 2014, MNRAS, 439, 1494

Bundy, K., Ellis, R. S., & Conselice, C. J. 2005, ApJ, 625, 621

Cohen, J. G., Cowie, L. L., Hogg, D. W., et al. 1996, ApJL, 471, L5

Conselice, C. J., Grogin, N. A., Jogee, S., et al. 2004, ApJL, 600, L139

Conselice, C. J., Blackburne, J. A., & Papovich, C. 2005, ApJ, 620, 564

Conselice, C. J. 2006, ApJ, 638, 686

Conselice, C. J., Rajgor, S., & Myers, R. 2008, MNRAS, 386, 909

Conselice, C. J. 2009, MNRAS, 399, L16
Conselice, C. J., Bluck, A. F. L., Ravindranath, S., et al. 2011, MNRAS, 417, 2770
de Vaucouleurs, G., de Vaucouleurs, A., Corwin, H. G., et al. 1991, Third Reference Catalogue of Bright Galaxies (New York: Springer)
Dahlen, T., Mobasher, B., Faber, S. M., et al. 2013, ApJ, 775, 93
Davis, M., Guhathakurta, P., Konidaris, N. P., et al. 2007, ApJL, 660, L1
Delgado-Serrano, R., Hammer, F., Yang, Y. B., et al. 2010, A&A, 509, A78
Dickinson, M., Dey, A., & Spinrad, H. 1995, Galaxies in the Young Universe (Berlin: Springer), 164
Dickinson, M., Hanley, C., Elston, R., et al. 2000, ApJ, 531, 624
Dressler, A., Oemler, A., Butcher, H. R., et al. 1994, ApJ, 430, 107
Driver, S. P., Windhorst, R. A., & Griffiths, R. E. 1995, ApJ, 453, 48
Dubinski, J., Mihos, J. C., & Hernquist, L. 1999, ApJ, 526, 607
Duncan, K., Conselice, C. J., Mortlock, A., et al. 2014, MNRAS, 444, 2960
Ellis, R. S. 1997, ARA&A, 35, 389
Elmegreen, B. G., Elmegreen, D. M., & Hirst, A. C. 2004, ApJ, 612, 191
Elmegreen, D. M., Elmegreen, B. G., Ravindranath, S., & Coe, D. A. 2007, ApJ, 658, 763
Erb, D. K., Steidel, C. C., Shapley, A. E., et al. 2006, ApJ, 646, 107
Ferguson, H. C., Dickinson, M., & Williams, R. 2000, ARA&A, 38, 667
Förster Schreiber, N. M., Genzel, R., Bouché, N., et al. 2009, ApJ, 706, 1364
Giavalisco, M., Ferguson, H. C., Koekemoer, A. M., et al. 2004, ApJL, 600, L93
Glazebrook, K. 2013, PASA, 30, e056
Griffiths, R. E., Ratnatunga, K. U., Neuschaefer, L. W., et al. 1994, ApJ, 437, 67
Grogin, N. A., Kocevski, D. D., Faber, S. M., et al. 2011, ApJS, 197, 35
Hambleton, K. M., Gibson, B. K., Brook, C. B., et al. 2011, MNRAS, 418, 801
Harrison, C. M., Johnson, H. L., Swinbank, A. M., et al. 2017, MNRAS, 467, 1965
Hernquist, L., & Katz, N. 1989, ApJS, 70, 419
Jogee, S., Barazza, F. D., Rix, H.-W., et al. 2004, ApJL, 615, L105
Koekemoer, A. M., Faber, S. M., Ferguson, H. C., et al. 2011, ApJS, 197, 36
Labbé, I., Rudnick, G., Franx, M., et al. 2003, ApJL, 591, L95
Lotz, J. M., Koekemoer, A., Coe, D., et al. 2017, ApJ, 837, 97
Law, D. R., Shapley, A. E., Steidel, C. C., et al. 2012, Natur, 487, 338
Lee, B., Giavalisco, M., Williams, C. C., et al. 2013, ApJ, 774, 47
Le Fèvre, O., Vettolani, G., Garilli, B., et al. 2005, A&A, 439, 845
Lotz, J. M., Jonsson, P., Cox, T. J., et al. 2008, MNRAS, 391, 1137
Lotz, J. M., Jonsson, P., Cox, T. J., et al. 2010a, MNRAS, 404, 575
Lotz, J. M., Jonsson, P., Cox, T. J., et al. 2010b, MNRAS, 404, 590
Mannucci, F., Cresci, G., Maiolino, R., et al. 2010, MNRAS, 408, 2115
Margalef-Bentabol, B., Conselice, C. J., Mortlock, A., et al. 2016, MNRAS, 461, 2728
Margalef-Bentabol, B., Conselice, C. J., Mortlock, A., et al. 2018, MNRAS, 473, 5370
Masters, K. L., Nichol, R. C., Hoyle, B., et al. 2011, MNRAS, 411, 2026
Melvin, T., Masters, K., Lintott, C., et al. 2014, MNRAS, 438, 2882
Mihos, J. C. 1995, ApJL, 438, L75
Mihos, J. C., & Hernquist, L. 1996, ApJ, 464, 641
Mortlock, A., Conselice, C. J., Hartley, W. G., et al. 2013, MNRAS, 433, 1185
Mortlock, A., Conselice, C. J., Hartley, W. G., et al. 2015, MNRAS, 447, 2
Price, S. H., Kriek, M., Barro, G., et al. 2020, ApJ, 894, 91

Schmidt, M. 1963, Natur, 197, 1040
Scoville, N., Abraham, R. G., Aussel, H., et al. 2007, ApJS, 172, 38
Shapley, A. E. 2011, ARA&A, 49, 525
Sheth, K., Elmegreen, D. M., Elmegreen, B. G., et al. 2008, ApJ, 675, 1141
Simmons, B. D., Melvin, T., Lintott, C., et al. 2014, MNRAS, 445, 3466
Snyder, G. F., Lotz, J. M., Rodriguez-Gomez, V., et al. 2017, MNRAS, 468, 207
Spinrad, H., Djorgovski, S., Marr, J., et al. 1985, PASP, 97, 932
Steidel, C. C., Pettini, M., & Hamilton, D. 1995, AJ, 110, 2519
Taylor-Mager, V. A., Conselice, C. J., Windhorst, R. A., et al. 2007, ApJ, 659, 162
Toomre, A., & Toomre, J. 1972, ApJ, 178, 623
van den Bergh, S., Abraham, R. G., Ellis, R. S., et al. 1996, AJ, 112, 359
Whitney, A., Conselice, C. J., Duncan, K., & Spitler, L. R. 2020, ApJ, 903, 14
Williams, R. E., Blacker, B., Dickinson, M., et al. 1996, AJ, 112, 1335
Williams, R. E., Baum, S., Bergeron, L. E., et al. 2000, AJ, 120, 2735
Williams, R. 2018, Hubble Deep Field and the Distant Universe (Bristol: IOP Publishing)
Windhorst, R. A., Schmidtke, P. C., Pascarelle, S. M., et al. 1994, AJ, 107, 930

Christopher J Conselice

Chapter 5

Physical Evolution of Galaxy Structure

This chapter introduces and reviews methods for uncovering the physical processes driving galaxy evolution based on structural observations, namely the ones that have been described in Chapter 3, using the data described in Chapter 4. This chapter will also focus on galaxy mergers—how we find them, the reliability of finding them, what the latest results from the different methods are for finding mergers and measuring merger rates.

This chapter will furthermore demonstrate how physical evolution is derived from the data, and explain how the measurement of galaxy mergers can be used to determine the merger history and rate, and therefore also the role of mergers in driving the formation of galaxies. As we have described earlier in this book, the change in galaxy structure includes not only the morphology and structure as measured, but also the sizes and stellar masses of galaxies through cosmic time. We also examine how the physical evolution of galaxies occurs and what are the driving mechanisms for it based on the changes in galaxy structure. This chapter also connects these structural observations to other observations of distant galaxies such as kinematic motions as observed through spectroscopy, including integral field units (IFUs). All of these measurements and properties add up to give a physical picture for how galaxy assembly is occurring over cosmic time.

Solving the problem of how and when the population of local galaxies formed and evolved into their physical and structural states is a major goal of astrophysics, and has been a fundamental one since galaxies were first known to be extragalactic. A large fraction of the stellar (and other forms of) mass in the universe are found within galaxies. Simply for this reason, understanding the origins of these systems is crucial for developing any ultimate theory of how structure, in a broad sense, in the universe developed and evolved.

Galaxy formation is at the forefront of the cosmological understanding of our universe since many major cosmological questions and parameters are well known now (Spergel et al. 2003). Thus even though we have a good understanding of the

doi:10.1088/2514-3433/abb602ch5

theory behind large-scale structure formation, we still do not have a good idea for how small-scale features such as galaxies themselves have formed and evolved. In fact, when comparing galaxy formation processes such as mergers with predictions from simulations it is often the case that even the most modern simulations still cannot produce correctly the merger history of galaxies in some instances (e.g., Bertone & Conselice 2009; Mundy et al. 2017; Duncan et al. 2019).

However, as described in this chapter there is progress on this front. In particular it is clear that an empirical way to measure galaxy evolution is preferred to a theoretical one. That is, with modern tools we can directly measure the processes responsible for galaxy formation—namely star formation and merging, which thereby gives us the full history of galaxy assembly that does not require us to rely on models of galaxies with fine-tuned physics.

5.1 Local Universe Benchmarks

Before we explore the distant universe it is important to take stock at what is the local universe's state for the types of galaxies we can see today and their distribution in properties. We use this information as our baseline, or benchmark, and view it philosophically as the ultimate final state of galaxy formation. This approach allows us to infer evolution for galaxies using mostly imaging. Imaging is an extremely powerful approach toward understanding this problem as spectroscopy, which is a more traditional way to obtain physical information for galaxies, is very difficult to obtain for many systems.

To see where we stand with galaxy formation, and the ways in which to uncover it in early phases of the universe it is important to be as quantitative as possible when describing the nearby galaxy population. For our purposes in this book, the nearby bright galaxy population is defined as galaxies brighter than $M_B < -20$, as these are the easiest galaxies to study and also are the most direct population we can compare with more distant galaxies, which have to be bright for us to detect. These nearby bright galaxies are becoming fairly well characterized through surveys such as the Sloan Digital Sky Survey (SDSS) and GAMA (e.g., Kauffmann et al. 2003; Bell et al. 2003; Kelvin et al. 2014). It appears that ~30% of the luminosity emitted in the nearby universe originates from early-type galaxies, with an even higher fraction of stellar mass found within this morphological type (e.g., Fukugita et al. 1998; Blanton et al. 2001; Hogg et al. 2002; Baldry & Glazebrook 2003; Conselice 2006a; Kelvin et al. 2014). These nearby bright galaxies are however clearly bimodal, containing a large population of old massive systems combined with less massive, on average, galaxies with somewhat younger stellar populations, such as spirals with a bulge and a disk (Kauffmann et al. 2003; Baldry & Glazebrook 2003). The entire morphological distributions of nearby galaxies using samples such as the "Third Reference Catalogue of Bright Galaxies" (de Vaucouleurs et al. 1991), show that about 75% of all nearby bright galaxies with $M_B < -20$ are spirals, while about 23% are ellipticals and the remainder are irregular galaxies (e.g., Marzke et al. 1994; Conselice 2006a; Figure 1.8).

A major question from this is answering: how and what time did this $z \sim 0$ local galaxy morphological distribution come into place? Furthermore, how are these processes of morphological assembly related to the development of stellar populations? We can answer these questions directly by studying in detail the structural and stellar mass evolution of galaxies at high redshift as seen with the Hubble Space Telescope (HST) using the type of data described in the previous chapter.

These questions have existed since galaxies were first imaged, especially whether their structures provided any clues to their formation. The Hubble sequence was thought to be an evolutionary one when it was initially created by Hubble based on work by James Jeans (1919). Within the Jeans theory this sequence was such that the ellipticals formed first, which then evolve into the spiral galaxies. As such, the ellipticals were known as “early types” and the spirals as “late types”—an outdated nomenclature that still is used (as it is as well, also incorrectly, for stars). The evolutionary connection of morphology to time is no longer thought to be valid, or at least is much too simplistic.

However, the structural criteria used by Hubble to classify systems were partially based on the relative ratio, as judged by eye, of the amount of light and sizes of the bulges and disks in spiral galaxies. It turns out that the disks of galaxies are usually dominated by star formation while bulges in the centers of these systems typically contain older stellar populations, such as is found for elliptical galaxies. These facts however were not understood at the time, and it was only shown decades later when the star formation rate for these systems could be measured (e.g., Kennicutt 1989). This means that the Hubble sequence is unlikely to be a simple evolutionary one with time, and to determine how this formation occurred we have to look back in time at the earliest galaxies and trace this evolution directly.

The simplest examination is simply just to ask what we see when we look back in the distant universe. As outlined in Chapter 4, the thought by some astronomers at the time before the Hubble Deep Field was that a deep picture of the distant universe would show a handful of relatively nearby galaxies, but that in the deeper and fainter parts of the image no galaxies would be seen. However, as have seen that is not at all what was found, and the Hubble Deep Field itself contains a few thousand galaxies, with the majority of these in the distant universe at $z > 0.5$. The next sections of this chapter explain what these observations are, and how they point in the direction of an evolution which we can infer from changes in the qualitative and quantitative properties of the galaxy population over time.

5.2 The Distant Universe

Extensive deep ground-based and HST imaging (e.g., Glazebrook et al. 1995; Driver et al. 1995; Madau et al. 1996; Brinchmann et al. 1998; Steidel et al. 1999; Conselice et al. 2003; Dickinson et al. 2003; Conselice et al. 2005; Mortlock et al. 2013) has progressively revealed much about the basic evolutionary trends in the structure of galaxies over time. In an early paper, Lilly et al. (1995) showed using an I-band selected survey of galaxies that the luminosity density, which can be inferred to match the star formation rate, increases with higher redshift out to $z \sim 1$. A more detailed analysis of a similar quantity by Madau et al. (1996), but this time using the rest-frame UV luminosity densities to $z \sim 4$, clearly showed that the star formation rate increases by up to a factor of 10 compared to the nearby universe up to high redshifts. This was the first qualitative indication that galaxy evolution had occurred gradually over time, and that it was not simply a very fast or very slow process.

At the same time, many of these star-forming galaxies were shown to be irregular or peculiar in appearance through HST imaging (e.g., Brinchmann et al. 1998). The third pillar of these studies is to examine the stellar mass evolution of these galaxies (e.g., Dickinson et al. 2003; Fontana et al. 2003; Glazebrook et al. 2004). These mass evolution studies, typically done in these early papers up to $z \sim 3$, found that only approximately 10% of modern stellar mass is formed by this redshift, which rises to ~50%–75% by $z \sim 1$. Thus, the bulk of stars in galaxies appears to form between these redshifts $z = 1$–3, occupying over 3.5 Gyr of cosmic time.

If we assume that these results are not biased by missing a large population of low surface brightness massive galaxies, or by a large cosmic variance, then already we can conclude that galaxy formation is a gradual process occurring over the age of the universe. It appears from this and more modern observations (e.g., Duncan et al. 2014; Mortlock et al. 2015) that most observations of stellar mass evolution roughly agree concerning when the mass in galaxies formed. The next stage is to determine how this formation occurred, as well as for ways in which the star formation history is driven over the same period of time (e.g., Madau & Dickinson 2014).

In general, there are at least three primary methods for instigating the assembly of galaxies: major mergers (mass ratios of 1:3 or lower), minor mergers (mass ratios of 1:3 or higher), in-situ star formation from gas which already exists in galaxies, and the accretion of gas from the intergalactic medium. Solving for which of these processes are occurring, as a function of time, requires examining the internal features of galaxies either through high resolution imaging or integral field spectroscopy.

Understanding this can be done through a number of ways, including the morphological evolution of galaxies, both through visual estimates as well as using quantitative indices such as CAS (Concentration, Asymmetry, Clumpiness) as outlined in Chapter 3. We can also combine this information with color, stellar mass, star formation, and luminosity information. In the future, machine learning techniques described in Chapter 8 will greatly advance this type of work.

5.3 Galaxy Morphological Evolution

As explained in previous chapters on the Hubble Deep Fields, analyses show that it is clear that galaxies are changing in morphology with time. Thus, it has been known for quite a while now that a significant fraction of all galaxies in the distant universe have irregular/peculiar morphologies. When this was first found it was thought that these galaxies may be signs of merger activity (e.g., Glazebrook et al. 1995; Driver et al. 1995; Abraham et al. 1996). These very early studies on galaxy structure were however limited and most of them did not have redshift information and simply quoted their results in terms of magnitude. Thus, these papers and studies could not trace the cosmic evolution of galaxy structure. Redshifts were used in later studies, such as Brinchmann & Ellis (2000), van den Bergh et al. (2001), and Kajisawa & Yamada (2001) who based on these observations argued that there is an evolution in galaxy morphology as a function of time. Specifically, van den Bergh et al. (2001) argued that very few ellipticals and spirals are found at redshifts $z > 1.5$ and other morphological properties change slowly at $z < 1.5$.

From this and other anecdotal evidence it became clear that galaxies transform from Hubble types in the nearby universe, which are mostly ellipticals and spirals, into more peculiar galaxies as we go to more distant times and higher redshifts. However, it is important to try to quantify this change with time. As such, we can examine how the morphologies of galaxies change as we probe more distant systems in deep HST imaging as recounted in Chapter 4. Because of the limitations of the filters on the HST we are only able to probe the rest-frame optical light of galaxies up to redshifts of $z \sim 3$. The results of this type of analysis was discussed in Chapter 4 and Figure 4.7. This figure shows a very basic examination of how morphologies change within the Hubble CANDELS fields, which probes the same rest-frame wavelengths in the optical up to this limit of $z \sim 3$.

Other work, such as Conselice et al. (2005) show that at higher redshifts a larger fraction of the total amount of stellar mass in all galaxies is attached to peculiars. At lower redshifts the amount of stellar mass found in peculiar galaxies gradually decreases, while the amount in ellipticals and spirals increases. This verifies on a longer timescale, the results of earlier work by Brinchmann & Ellis (2000) who see a similar type of evolution within galaxies at $z < 1$. Another finding of Conselice et al. (2005) is that at $z \sim 1.5$ the co-moving densities, co-moving B-band luminosities and co-moving stellar mass densities are nearly equal for peculiars, disks, and ellipticals. This is also the redshift where disks and ellipticals first appear morphologically in some abundance. At redshifts $z > 1.5$ peculiar galaxies, which, as we shall argue, are likely mostly galaxies in some merger state, increasingly dominate the number, luminosity, and mass densities of galaxies up to redshifts $z \sim 3$.

Whatever effects are driving galaxies to become peculiar at these higher redshifts are therefore the dominant mode for forming the most massive galaxies at $z > 1.5$. As we argue below, these are likely to be major mergers while minor mergers and gas accretion from the intergalactic medium are the likely cause of star formation and stellar mass assembly at lower redshifts.

5.4 Morphological k-Corrections

When we observe a galaxy in the distant universe, several things change within that galaxy. Because galaxies are not just more distant, but appear to be moving away at a high radial velocity due to the Hubble expansion of the universe, there are several effects both in terms of the amount of light seen, as well as the size, surface brightness and morphology (see Section 3.5).

In cosmological calculations, the angular size distance decreases up to redshifts of near $z \sim 1.5$ and then increases at higher redshifts. That is, galaxies at the highest redshift have an angular size that increases with redshift past $z \sim 1.5$. The luminosity distance declines much faster with redshift, such that the surface brightness of galaxies declines as $(1 + z)^{-4}$, meaning that many distant galaxies are difficult to detect and study and some may not be detected at all (e.g., Whitney et al. 2019). Due to the Doppler effect this also implies that the rest-frame wavelength we observe in these galaxies scales with the observed wavelength by the formula:

$$\lambda_{\text{rest}} = \frac{\lambda_{\text{observed}}}{(1 + z)}. \tag{5.1}$$

What this means is that using the HST filter set, even with WFC3 which goes into the infrared, it is impossible to probe the rest-frame optical light from galaxies past $z \sim 2.5$. In Figure 3.8 we plot the rest-frame wavelengths probed using common HST fields (from Conselice et al. 2011a).

From this we can conclude that all HST observations of more distant galaxies than this limit are being imaged in the rest-frame ultraviolet, where it is possible that galaxy structures are very different from what they are in the rest-frame optical (e.g., Windhorst et al. 2002). The difference between a galaxy's morphology at different wavelengths is called the morphological k-correction, named as such to be similar to the k-correction itself, which the amount of magnitude corrections for galaxies of various types observed at higher redshifts (Kinney et al. 1996). This can be extended to account for a quantitative morphological k-correction, which is the difference of a measured index in terms of the measurements of its parameters (e.g., Conselice et al. 2005; Chapter 3).

We know that the change in morphology we see in galaxies at higher redshifts is not very likely a result of this so-called morphological k-correction. This is especially the case for those galaxies at high redshift where we have the ability to probe the rest-frame optical light. Before WFC3 data was widely available, we had limited NIR imaging from the NICMOS camera on Hubble. Using this, Conselice et al. (2003) utilized imaging of the Hubble Deep Field (HDF) to argued that the quantitative morphological k-correction is not significant for galaxies when observed at higher redshifts. The reason for this is that galaxies that are dominated by star formation, as most galaxies in deep HST imaging are, look very similar in the UV and rest-frame optical (e.g., Conselice et al. 2000). In Figure 5.1 we show galaxies in the distant universe from $z = 1.5$ to $z = 2.5$ probed at HST filters from the rest-frame optical to the rest-frame UV. While the signal to noise of these observations does decrease at shorter wavelengths, often the galaxies retain their overall morphology and structure.

5.5 Size Evolution

One of the most basic features of galaxies is their sizes, or rather how large they are. This is a fundamental feature and it has been noted for sometime (e.g., Trujillo et al. 2007) that galaxies at the same stellar masses are smaller in the more distant universe. What this means is that galaxies have evolved more in size than they have in mass from at least $z < 3$. We can easily see this by plotting this change in size as a function of redshift for galaxies with high stellar masses (Figure 5.2).

The evolution seen in size is also not a small amount of change, but can be a factor of 5–10 in terms of radius with no obvious growth in the stellar mass. One

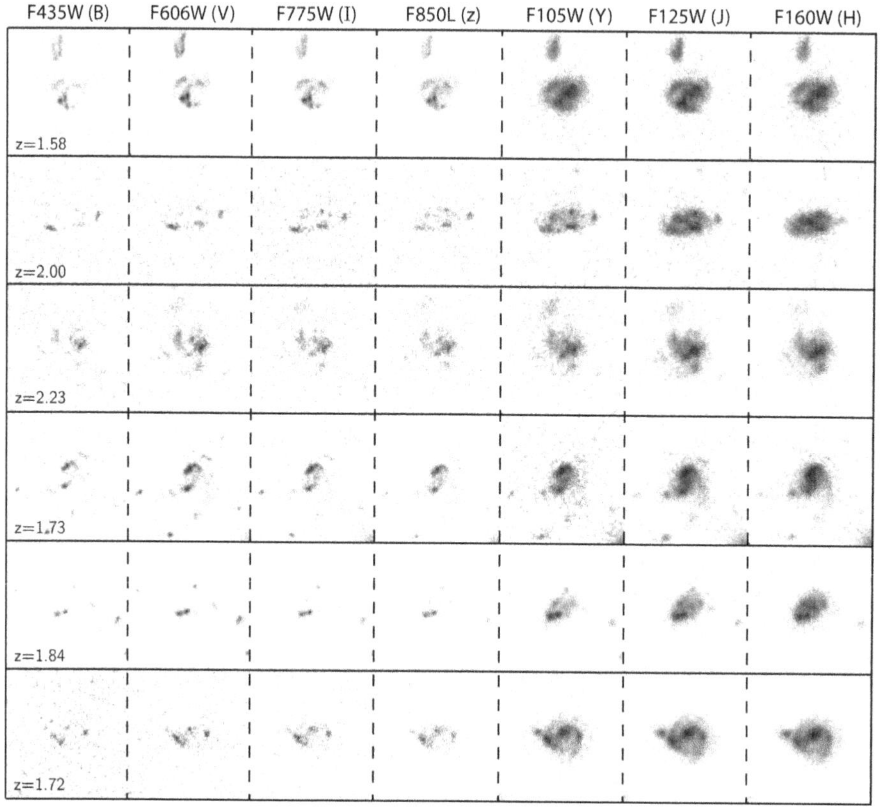

Figure 5.1. This figure shows the morphological changes in galaxies observed at high redshift as a function of observed wavelength. Shown are images of galaxies in all the filters available within the CANDELS fields, demonstrating that some galaxies can look quite different at shorter wavelengths where the signal to noise is lower and the stellar populations probed are focused on younger stars within these systems. At higher redshifts it is often the case that galaxies are dominated by star formation and the morphology of the systems is fairly similar in the rest-frame UV and rest-frame optical light (courtesy of Dale Kocevski).

example of this evolution is shown in Figure 5.2 using the results from the GOODS NICMOS Survey (GNS) and the POWIR surveys studies of galaxy evolution (Buitrago et al. 2008). It is clear from this, and later studies that expanded this to higher redshifts (e.g., Bluck et al. 2012; Whitney et al. 2019) that galaxies get progressively smaller at higher redshifts. The origin of this is still somewhat of a mystery, although it does appear that a large part of the evolution can be accounted for by minor mergers (e.g., Bluck et al. 2012). The reason that major mergers are not a possible solution to this size problem is that from simulations when two galaxies of similar mass merge, the size goes up roughly at the same rate as the mass increase. This means that the size versus mass relation is preserved after the merger, and thus this is not a solution for explaining how sizes can grow while mass does not. However, minor mergers are able to increase the size of a galaxy by building its outer portions with multiple merging events, thereby increasing both the Petrosian radii in the outer portions, as well as the effective radii (e.g., Whitney et al. 2019). If the

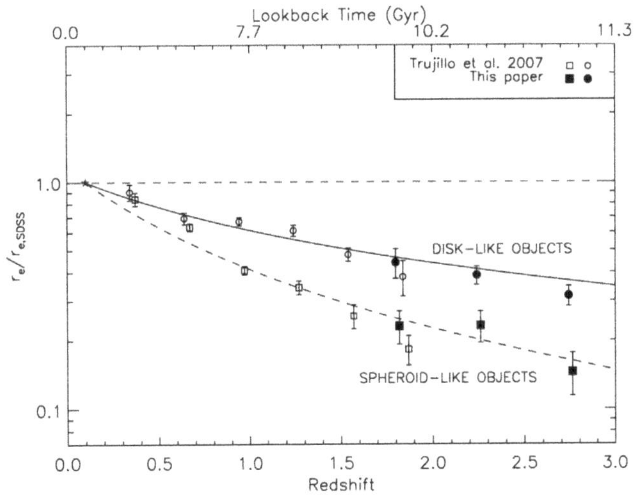

Figure 5.2. Size evolution of massive galaxies as observed in deep fields. As can be seen, there is a progression of obtaining smaller sizes for galaxies of a given mass the further back in redshift we observe these systems. For these massive systems we can see that the typical size is over a factor of five times smaller than similar mass galaxies seen in the local universe. Reproduced from Buitrago et al. (2008).

number of major mergers is indeed what is observed in data, then this is enough to account for this size evolution (Bluck et al. 2012).

5.6 The Galaxy Merger History

In the previous chapter we described HST observations of deep fields and how they reveal that galaxies are more peculiar when they are seen at higher redshifts. This peculiarity in the galaxy population declines as we go to lower redshifts. Chapter 3 included a discussion for how to use quantitative morphological and structural indices to determine properties of galaxies, such as their merger rates. In this section we utilize these observations to determine what are the merger rates of galaxies as a function of redshift, and ultimately what is the role of mergers in forming galaxies. In many ways, this is the most powerful application of galaxy structural analysis applied to imaging data we can carry out at the present time.

5.6.1 Defining Mergers and Merging

Before describing the background of galaxy mergers and how they have evolved throughout the universe it is critical to explain what we mean by galaxy mergers. Traditionally, there is some confusion within the literature concerning measuring the merger history, and thus what appears to be conflicting results on the evolution of mergers are the result of inconsistent definitions being used.

Before we explain the observed merger history it is important to obtain an overview of what we mean by the merger process for two galaxies. Within this discussion we assume that galaxies have already formed at least some stellar mass before they merge, or else there would not be galaxies to merge together. From this notion the basic idea is that when galaxies have formed their first masses, which can be very low at the level of ultra dwarf galaxies, they begin to merge. As the universe is so dense at early times, these galaxies at the start of the universe will likely collide and form more massive galaxies through the attraction of gravity. Due to this gravitational attraction these systems eventually collide and coalesce into a new galaxy—growing systems which become more massive over time. This process will continue to go on and on with this new galaxy smashing into and merging with another galaxy, and then another, and so on through cosmic time.

There are two main ways of finding interacting and merging galaxies that we will discuss throughout this chapter. The first are the classical mergers where the system looks peculiar (Figure 1.6) and the second type are those systems which are close together on the sky in "pairs" (see Figure 3.5). In the nearby universe merging galaxies are relatively rare, making up just 1% of all galaxies, with lower mass galaxies more likely to be undergoing a merger (Casteels et al. 2014).

When should we expect mergers to occur? Merging galaxies are more likely to be found when and where galaxies are closer together, such as the group environment and at higher redshifts when the universe is more dense. We discuss both of these situations below.

The most common type of mergers studied to date are the so-called major mergers. Major mergers are defined as merging systems whereby the mass ratio varies between equal mass (1:1) mergers, down to either a third or fourth of the more massive galaxy, relative to the lower-mass system (1:3) or (1:4). These major mergers dominate the study of galaxies simply because they are the easiest to find and study. This is due to the fact that when a major merger occurs it is between systems pre-merger which are of nearly equal mass. When this occurs and the galaxies are nearly similar in stellar mass then the effects on the galaxy's structure are easiest to see. The other type of merger—known as minor mergers, are when the stellar mass ratio is less than 1:3, often do not produce morphological or structural changes (see Chapter 4, Section 4.6.1) and often have to be located based on finding systems in pairs. Because of this, it is more difficult to find minor mergers since the secondary (i.e., lower mass) galaxy in the merger is usually much fainter than the primary. This is such that pairs with small mass ratios are difficult to detect (e.g., Duncan et al. 2019).

There is another type of merger, which is very difficult to detect which are called micro-mergers. These are mergers which occur with very low mass galaxies—systems of about 1/100th or less of the mass of the central galaxy. These mergers probably coincide with the addition of extensive amounts of cold gas which can furthermore increase the star formation rate of the galaxy. Because very little is known about these galaxies, and they have very little morphological or structural effects, we will not consider these mergers in much detail in this chapter.

It is also important to stress that the selection method of finding mergers is also important. If mergers are selected based on a stellar mass cut or a luminosity cut then different fractions and rates will be derived (e.g., Man et al. 2016). It is important to realize that this in no way implies that the merger fraction is inconsistent between studies, but that by selecting galaxies in different ways we are selecting the underlying galaxy population differently which leads to different merger fractions. The merger fraction can vary significantly with galaxy property (e.g., Casteels et al. 2014), thus the choice of how mergers are defined is critical for producing consistent merger histories.

5.6.2 Galaxy Merger Fraction Evolution

As explained in Chapter 3 in detail, we are able to determine mergers from galaxies by using structural indices such as the asymmetry and clumpiness parameters. When this methodology is applied to distant galaxies it is easy to show that the derived merger fraction increases significantly as we probe higher redshifts.

Historically, the merger fraction was first quantified in this way in Conselice et al. (2003) who demonstrated that the peak merger fraction at $z = 2.5$ was about 40%. Conselice et al. (2003) also parameterized the merger fraction increase as a power-law which goes as $(1 + z)^m$, where m is the power-law index for the increase of the fraction of mergers with time, and is generally positive. Early studies such as this on the merger history derived values of $m \sim 3$.

This formulaic way for quantifying the merger history was first introduced by Carlberg (1991). This was originally intended as a way to measure the increase in the merger history for quantifying cosmological parameters of the universe. It was thought that this value of m correlates with the deaccleration parameter. The full formula for this merger history increase with redshift can be expressed as:

$$f_{\rm m} = f_0 \times (1 + z)^m, \tag{5.2}$$

where the value f_0 is the merger fraction at $z = 0$ and m as described is the merger power-law index. This formula fits very well the increase in the merger fraction up to $z \sim 2$. It turns out that this merger fraction may turn over at higher redshifts such that it stops increasing so quickly with redshift (e.g., Conselice et al. 2008). Furthermore, the value of the power-law index m depends highly on the value for the $z = 0$ merger fraction value, f_0. If the redshift $z \sim 0$ merger fraction is not known, or is held constant at a too high or too low of a value, then the fitted value of m can be measured incorrectly.

The use of this fitting form, and investigation of the merger history was however initially only done within the narrow field of the Hubble Deep Field North using NICMOS data. From Conselice et al. (2003) onward, various other studies have measured this merger history. First, for lower redshifts, it was found that the merger history up to $z \sim 1$ in the COSMOS field shows that the merger fraction measured with morphology continues to increase with redshift up to this limit and can be fit by a power-law (Kartaltepe et al. 2007; Conselice et al. 2009).

Early studies measuring the merger fraction through galaxies in pairs also found a very steep increase in the merger fraction up to $z \sim 1$, with a fitted index of $m = 2.8 \pm 0.9$ for a luminosity selected sample up to $z \sim 0.4$ (Patton et al. 1997). This fitted value of m was later calculated by Patton et al. (2002) as $m = 2.3 \pm 0.7$ within the CNOC2 redshift survey up to $z \sim 0.55$. Other studies such as Le Févre

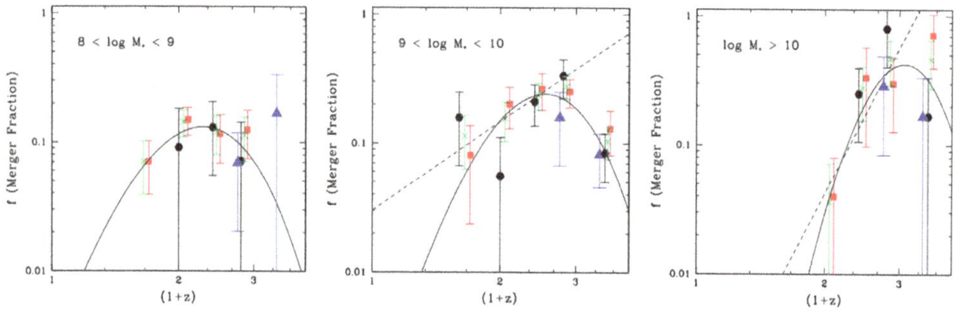

Figure 5.3. This figure shows the most up to date version of the merger history using CAS morphology (Conselice et al. 2011a, 2011b) for the deepest Hubble Space Telescope imaging using the Hubble Deep Field and Ultra Deep Field, including infrared imaging. It shows the merger fraction as a function of redshift and stellar mass for the UDF WFC3 sample (blue triangles) compared with data from the NICMOS HDF-N (red squares), the ACS UDF (black circles) and the combination of the NICMOS and ACS data (green crosses). These merger fractions are plotted at the mass limits of $8 < \log M_* < 9$, $9 < \log M_* < 10$ and $\log M_* > 10$. The green crosses are the evolution of the merger fraction for the combined UDF and HDF-N sample.

et al. (2000) measured the merger fraction using 285 galaxies in the CFRS and LDSS surveys up to $z \sim 1$, and who found that the best fitting power-law index was even steeper with $m = 3.2 \pm 0.6$, although this possibly lowers to $m = 2.7 \pm 0.6$ when considering selection effects. Later studies found similar fitted values, with $m = 1.5 \pm 0.7$ for brighter galaxies in the VVDS (de Ravel et al. 2009), and $m = 3.1 \pm 0.1$ for pairs up to $z \sim 1$ in the COSMOS field (Kartaltepe et al. 2007). These studies are for relatively nearby galaxies at $z < 1$. At higher redshifts, Bluck et al. (2009) fit a power-law of $m = 3.0 \pm 0.4$ for galaxies with $M_* > 10^{11}\ M_\odot$ using pairs of very massive galaxies from the GOODS NICMOS Survey. This variation is not as significant as it may seem at first, as each of these studies used different criteria for finding mergers. This variation can also change depending on to what redshift these fractions were fit to, as well as different methods in which the mergers were selected.

Within morphological based studies using the CAS parameters, Conselice et al. (2003) found a high m index of $m \sim 4$ for massive galaxies with $M_* > 10^{10}\ M_\odot$, with lower m values around $m \sim 1 - 2$ for lower mass galaxies (Figure 5.3). Overall, this is quite similar to what was inferred by Le Févre et al. (2000) who examined the merger history of visually disturbed galaxies in deep Hubble imaging. Similar merger fractions using CAS were found by Conselice et al. (2008) within the Hubble Ultra Deep Field data. When examining the data it is clear that Conselice et al. (2003) and later studies showed that the merger fraction slope m varies significantly depending on what redshift limit is used, and whether stellar mass or luminosity cuts are applied to the selected sample. In a CAS based study of $z < 1$ galaxies from the COSMOS survey, Conselice et al. (2009) found an index of $m = 2.3 \pm 0.4$ for galaxies with stellar masses of $M_* > 10^{10}\ M_\odot$. While Lotz et al. (2008) using the Gini/M_{20} formalism found that the merger fraction is roughly constant with redshift between $z \sim 0.2$–1.2, with at best a minor increase in the merger fraction with redshift such that $m = 0.23 \pm 1.03$ using the Gini/M_{20} methods. However, as outlined in Chapter 4, the

Gini/M_{20} methodology is very sensitive to minor mergers, and once these effects are considered the fitted merger fraction slope using this methodology increases to $m = 2$ up to $z \sim 1.2$ (Lotz et al. 2010). Other methods for calculating the merger or interaction rate of galaxies, including looking for ringed galaxies (D'Onghia et al. 2008), found similar fitted merger fraction indices with $m \sim 3$.

An interesting feature of the merger history as measured using CAS parameters, which has permitted a higher redshift exploration of this history, is that the merger history tends to peak at $z \sim 2$ for massive galaxies at $M_* > 10^{10}\ M_\odot$ with values of $f_m \sim 0.3 - 0.4$ (Figure 5.3). After this redshift, the CAS derived merger fraction declines when going to higher redshifts. This is also now found when examining the merger fraction measured through using galaxy pairs. One way to represent this behavior is to parameterize the merger fraction evolution as a combined power-law and exponential; this is a form of the Euler gamma function (e.g., Conselice et al. 2008; Duncan et al. 2019) whose equation is given by:

$$f_m = \alpha \times (1 + z)^m \exp[\beta(1 + z)]. \tag{5.3}$$

The local $z = 0$ merger fraction in this formalism is calculated by $f_m(0) = \alpha \exp(\beta)$, and the merger peak is located at $z_{peak} = -(1 + m/\beta)$. These parameters are given in various papers including Conselice et al. (2008), Mundy et al. (2017), and Duncan et al. (2019). Examples of this merger turnover within deep HST imaging is shown in Figure 5.3.

Another way to determine the merger fraction evolution is to search for galaxies in pairs. This methodology is in someways more advanced than the morphological approach, at the time of writing, and ultimately we want to be able to reproduce the merger history using both approaches and for them to be consistent, which appears to be the case (Ferreira et al. 2020). The most up to date merger fraction history of galaxies using pairs is shown in Figure 5.4, as measured up to $z \sim 6$. These pair fractions are slightly different from those measured from the structural analyses. There is no obvious turn over at higher redshifts, but there is not as rapid an increase in the pair fraction at redshifts higher than $z \sim 2$, particularly for the lower mass galaxies in the upper panel of Figure 5.4. Figure 5.4 also shows that in general we are more likely to find the most massive galaxies, with $\log M_* > 10.3$, involved in mergers at higher redshifts compared with their counterparts at lower redshifts.

These measured pair fractions give us a good idea of the fraction of galaxies undergoing mergers occurring within a given galaxy population as a function of time or redshift. However, these values in themselves do not tell us how mergers are affecting the formation of galaxies. This is because the timescale in which these mergers occur is largely unknown from observations and have to often be inferred from dynamical friction calculations or N-body models as explained in Chapter 4. Thus to make advances, we need to understand these timescales; without this information we do not know how long a galaxy is in the merger phase we observe it in, and therefore we cannot calibrate the observed fraction into an inferred rate. In a real sense, these merger fractions are in the absence of any timescale information simply an observation outside of proper context to understand its implications.

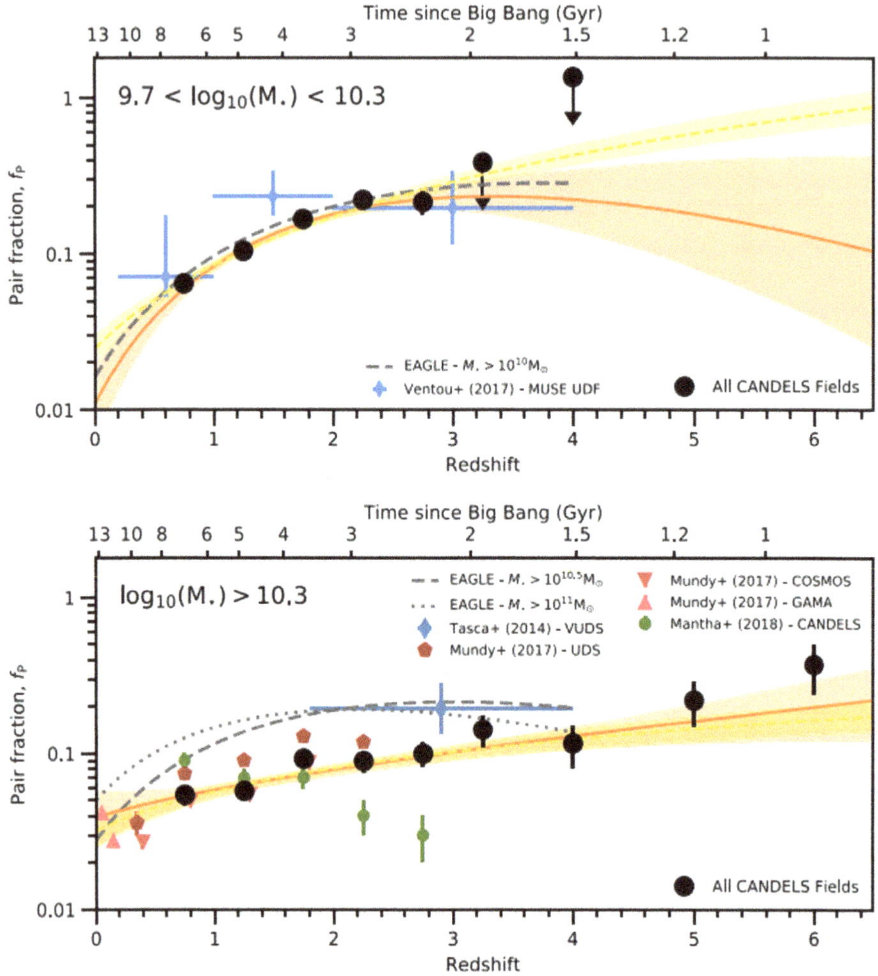

Figure 5.4. The measured pair fraction of galaxies up to $z \sim 6$. This plot shows our latest estimate, at the time of writing, of how the merger history changes as a function of time. The black points are from deep Hubble Space Telescope imaging while the other colored points are from large ground-based surveys, including Mundy et al. (2017). As can be seen, the pair fraction continues to increase at higher redshifts up until $z \sim 6$, where it reaches a value of ~50%. Also shown on this figure are model comparisons from the Eagle simulation, which generally find a higher pair fraction than what is observed. Reproduced from Duncan et al. (2019).

Although when the merger fractions can be interpreted with a reliable timescale it gives us the merger rate—which is the topic of the next section of this chapter.

5.6.3 Galaxy Merger Rates

What we ultimately want to know is the galaxy merger rate for understanding how mergers influenced the formation of galaxies and their constituents. It is basically the number of mergers per unit time within a given galaxy selection. Merger rates tell us how many mergers occur in a population over a given amount of time, and thus also how much mass is accumulated by these mergers over this time-span. The galaxy merger rate is however a partially theoretically derived quantity based on the timescales for galaxies to merge, and the measured merger fractions discussed in the previous section.

The timescales for galaxy mergers are quite different for galaxies in pairs versus galaxies which are morphologically disturbed. Galaxies which are separated by some apparent distance in a pair have merger timescales that can be calculated using the dynamical friction timescale (Section 4.6.2). This dynamical friction timescale depends on the masses of the systems, such that the more massive the merging galaxy relative to the central system the shorter the time period for the merger to happen.

For the pairs in Figure 5.4 the merger timescale ($\tau_{\rm m}$) we use to calculate the merger time is by using the dynamical friction calculation for galaxies with separations of ~20–30 kpc. Alternatively, this timescale can be investigated using N-body models of mergers, which give similar answers (e.g., Lotz et al. 2008). When we calculate this we find that the merger timescale for pairs varies between $\tau_{\rm m} = 0.5$–1 Gyr. This calculation, and ones similar to it, however assumes a few things. The most obvious is that we are assuming that these distant galaxies are dominated by dark matter, as local systems are. This is a reasonable assumption as we know that galaxies in the distant universe have a dark matter to seller mass ratio similar to what we find in local galaxies (e.g., Conselice et al. 2018). From this we can tentatively conclude that the dynamical friction formula is applicable in the distant universe. The merger rate is then defined as:

$$\Gamma(M_*, z) = \frac{f_{\rm m}}{\tau_{\rm m}}. \tag{5.4}$$

When we use galaxy structures such as those calculated with the CAS system to locate which peculiars are indeed merging galaxies to derive merger fractions, we have to use different timescales. When we use morphology and structure to measure the merger rates we have to use simulations to derive these values, as observationally measuring the merger process over ~1 Gyr years to determine how long mergers occur is very difficult (although see Conselice 2009). We thus have to infer from simulations what the timescale $\tau_{\rm m}$ is for these systems. We cannot use dynamical fiction timescales for this calculation simply because the merger has already occurred, and the resulting system is dynamically relaxing at this stage. We calculate these timescales through using simulations of the merger process as described in Chapter 4. Using these simulations also tells us which types of mergers produce measurable structures which can then be converted into a timescale.

This approach of calculating merger timescales from structure was first carried out by Conselice (2006b) who examined pairs of galaxies merging in N-body models. In this paper, Conselice (2006b) showed that the timescale for major mergers using asymmetric galaxies is around 0.5–1 Gyr. This is similar to what is found for galaxies with a pair separation of about 30 kpc. Furthermore, it can be shown through simulations that high values of the asymmetry index locate systems which have mergers with a merger ratio greater than ~1/3. Mergers with lower mass ratios do not produce a high enough asymmetry to be seen as a merger (as shown in Figure 4.12). Therefore, the mergers measured with CAS are those which are undergoing major mergers. This fact demonstrates that the asymmetry index is a robust method of finding major merging galaxies, as opposed to other indices such as the Gini/M_{20} method, whereby an unknown mix of minor and major mergers are identified (e.g., Lotz et al. 2010).

These analyses did not include galaxies undergoing star formation, or include other galaxy structures that can mimic asymmetries such as dust lanes. Since then, more detailed simulations incorporating star formation and other effects have been employed (e.g., Lotz et al. 2010), but the main results found by Conselice (2006b) concerning the timescales for merging still hold (see Chapter 4).

5.6.4 Number of Merger Events at $z < 6$

Resulting major merger rates from these calculations are shown in Figure 5.5 using the pair methodology up to $z \sim 6$. From this figure it can be seen that the derived merger rate increases quite rapidly until the highest redshifts where this can be probed. At $z \sim 6$ the merger rate is about 10 mergers per Gyr, demonstrating that there is on average a merger every 100 Myr at these redshifts for the most massive galaxies. This is currently the highest redshifts where we can measure these features currently. Deeper explorations in mass and redshift will require new telescopes such as the James Webb Space Telescope (JWST) (Chapter 8). This very rapid merger rate at early times implies that merging is an important process for forming galaxies when the bulk of its stellar mass is being assembled. When we are able to measure the timescale for mergers and thus the merger rate, we can then calculate how many mergers typically galaxies must have undergone through time. The formula for this calculation is given by:

$$\mathrm{N}_* = \int_{t_1}^{t_2} \int_{M_1}^{M_2} \frac{f_\mathrm{m}(z, M_*)}{\tau_\mathrm{m}(M_*)} dM_* dz, \tag{5.5}$$

where N_* is the number of mergers which have occurred between two masses M_1 and M_2 over the timescale from t_1 to t_2. Based on this formalism, Conselice (2006b) measured that a typical massive galaxy undergoes $4.4^{+1.6}_{-0.9}$ mergers from $z \sim 3$ to $z \sim 0$. Later calculations have derived slightly different numbers of major mergers over this timescale with typically lower values given lower merger fractions measured in later work (e.g., Duncan et al. 2019). Duncan et al. (2019) used this formalism to show that over the redshifts $z \sim 0$–6 there are on average a few major mergers for galaxies at large masses. The merger history is very similar if instead of selecting with stellar mass a co-moving volume selected sample of galaxies is used (e.g., Duncan et al. 2019). However, in addition to the number of mergers we are also very interested in knowing how much mass is added to galaxies through the merger process—i.e., measuring the role of mergers in the galaxy formation process.

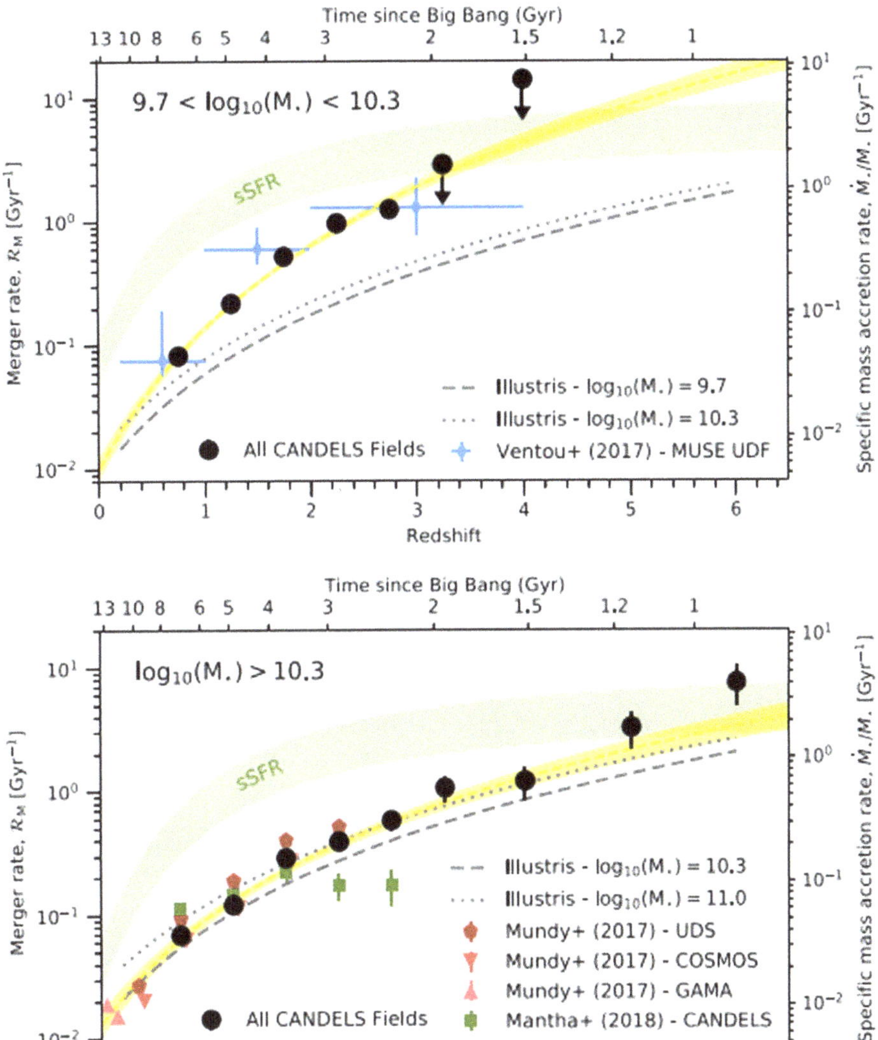

Figure 5.5. The merger rate (number of mergers occurring per galaxy per unit time) as a function of redshift (left hand side) for galaxies in pairs up to $z \sim 6$. The right-hand side shows the specific mass accretion rate, the amount of stellar mass brought into a galaxy due to the merger process. The dashed and dotted lines show the predictions for these merger rates within the Illustris simulation as described in Duncan et al. (2019). The shaded green curve shows the corresponding specific star formation rate (sSFR) for galaxies at the same mass selection as for the mergers as a function of redshift. Results from previous work, including Mundy et al. (2017) and Mantha et al. (2018) are also shown. Reproduced from Duncan et al. (2019). © 2019. The American Astronomical Society. All rights reserved.

This does not include the amount of mass added to galaxies through triggered star formation due to mergers, a topic we discuss in Chapter 6.

5.7 Mass Assembly Through Merging

The merger rate gives us some idea about that number of mergers occurring as a function of time, but this by itself does not reveal how mergers are influencing the formation of mass within galaxies over this cosmic time. For this we need to calculate how much stellar mass is added to these galaxies over time due to these mergers. Ultimately we would like to know how much gas mass is added to these galaxies or the amount of dark matter, but these quantities are too hard to measure currently in distant galaxies, and thus we are constrained by only being able to examine the amount of stellar mass, which can more easily measured with galaxy photometry.

Before describing the process for how mass is added to galaxies due to the merger process we review the very basics for how we think galaxies are assembling. There are two main ways in which we think galaxy formation occurs. These are the process of converting gas into stars through star formation while the second is the assembly of pre-existing galaxies into a new galaxy through the merger process already described. Luckily, we already have a good observational idea of the ongoing star formation rate for galaxies as a function of time. As shown in many studies the star formation rate density for galaxies integrated over all masses increases when we look back in time, and has a peak rate at about $z = 2$ (Madau & Dickinson 2014), which roughly agrees with the evolution of the merger rate (Conselice 2014). What we need to do is however determine how much stellar mass is added to galaxies during this merger process and compare it to the star formation rate to conclude which process is the more dominant one with time. Ultimately comparing these two methods as a function of galaxy mass and time would be a powerful way to determine the relative importance of difference processes of galaxy formation.

First, we examine the formalism for calculating the amount of stellar mass brought in through mergers. The total amount of stellar mass added into a galaxy is a double integral over the redshift range in which we are interested in determining the mass build up. For our purposes we will call this range z_1 and z_2, which corresponds to look-back times of t_1 and t_2, respectively. This integral is also over the stellar masses we probe whose range is denoted by M_1 to M_2. For deep surveys such as the GOODS NICMOS Survey (e.g., Conselice et al. 2011b; Bluck et al. 2012) or CANDELS (e.g., Mortlock et al. 2015) the mass range probed goes down to $M_* = 10^{9.5}\,M_\odot$. Overall we can express this additional mass gained through the formula:

$$\mathrm{M}_{*,\mathrm{M}} = \int_{t_1}^{t_2} \int_{M_1}^{M_2} M_* \times \frac{f_{\mathrm{m}}(z, M_*)}{\tau_{\mathrm{m}}(M_*)} dM_* dz, \tag{5.6}$$

where $\tau_{\mathrm{m}}(\mathrm{M}*)$ is the merger timescale, where we have added in this formalism a dependence on the stellar mass. This is because results show that there is a dependence of the merger history on the stellar mass of the merging pair (Bluck et al. 2012). From this we can calculate the total integrated amount of stellar mass assembled through merging. We express this as a ratio such as when we divide it by the original mass of the galaxy we obtain a dimensionless quantity, $\dot{M}/M = M_{*,M}/M_{*}(0)$. For the GNS we find a value of $M_{*,M}/M_{*}(0) = 0.56 \pm 0.15$ (Conselice et al. 2013). We can use this to get an idea of how much additional mass from the initial mass was added due to the merger process (where $M_{*}(0)$ is the initial average stellar mass).

The right-hand side of the plots in Figure 5.5 from Duncan et al. (2019) show the amount of stellar mass accreted onto galaxies due to the merger process, in terms of the specific mass accretion rate plotted as $\dot{M}/M$, where $\dot{M}$ is the amount of mass brought in through mergers. The relative form of the evolution of this merger mass accretion rate traces and mimics the merger rate itself. As can be seen, the specific mass accretion rate is very high at high redshifts, whereby the bulk of matter in galaxies which forms from mergers does so. We can also compare this mass addition from merging to the specific star formation rate (sSFR) for the same galaxies; the sSFR is an entirely analogous fraction of mass formed through star formation divided by the initial mass of the galaxy. As can be seen, the star formation rate is much more important for building up the mass of galaxies at lower redshifts, but the specific mass accretion rate is similar, or higher, at $z > 5$. From this we can conclude that at least major mergers are a dominant way in which galaxies are forming their mass at very early times in the universe's history. This is however not necessarily the case at lower redshifts where only up to 40% of the mass in galaxies is established through major and minor mergers (e.g., Ownsworth et al. 2014).

Minor mergers are also found to be quite common during this epoch, but are much harder to find and quantify. This is because these mergers are too difficult to locate—their effect on the morphology of galaxies is hard to see, as galaxies that are merging with much smaller galaxies do not change very much in terms of their morphological structure (e.g., Conselice 2006b; Lotz et al. 2010). However, we can use galaxy pairs to trace this process (e.g., Bluck et al. 2012; Duncan et al. 2019). What we generally find is that minor mergers are about as common as major mergers in terms of adding mass into galaxies. There are more minor mergers over the epochs of interest for galaxy formation, but their total addition of mass is similar to major mergers (e.g., Ownsworth et al. 2016).

The higher number of minor mergers is furthermore a potential solution to the problem of finding massive galaxies with small sizes at high redshifts which become larger at lower redshifts. As described in Section 5.5 the effective radii of massive galaxies at $z > 0.5$ are up to a factor of five times smaller than galaxies of similar masses in the local universe (e.g., Buitrago et al. 2008; Trujillo et al. 2007; Weinzirl et al. 2011). If we examine the number of minor mergers we observe from $z = 3$ down to lower redshifts it is possible that these mergers provide enough stellar mass in the outer parts of these galaxies such that their measured sizes grow by up to a factor of five (Bluck et al. 2012; Ownsworth et al. 2016; Whitney et al. 2019). This is therefore a likely solution to the problem of how galaxies can be so compact and with high mass at high redshift.

5.8 Gas Accretion From the Intergalactic Medium

In addition to the stellar mass we also want to understand and calculate the total baryonic mass placed into galaxies due to the merger and other accretion processes. Baryons are the stars and gas which will usually assemble together in a galaxy through a hierarchical way. In addition to stars, we also need to account for how much cold gas mass, which can trigger star formation, is brought into these systems through the merger process. We calculate this by integrating the amount of gas brought in through these merging systems. Often the problem is however that we do not know or have a good way to measure the gas masses for the majority of galaxies that can be observed in the distant universe. To do this we thus use an empirical fit to the relationship between the gas mass fraction, $\mu_{\rm gas}$, and the stellar mass found at $z = 2 - 3$ (e.g., Conselice et al. 2013).

It turns out that the lower mass galaxies contribute the bulk of the gaseous mass acquired from minor mergers, whereas most of the stellar mass added to galaxies in mergers arises from major mergers. We can then use this relation to calculate how much gas mass is added due to merging, also in terms of the ratio $M_{g,M}/M_*(0) = M_{*,M}/M_*(0) \times f_g$. For the GOODS NICMOS Survey sample the value for this for galaxies over $z = 1–3$ is found to be $= 0.57 \pm 0.15$. It also appears to be the case that over the redshift interval $z = 0–3$ major and minor mergers are roughly similar in importance for building up the baryonic masses of galaxies, with the minor mergers contributing gas, while major mergers contribute mostly stars. However, this gas mass brought in is not enough to account for all the star formation seen in these massive galaxies over the epoch $z < 3$ and thus it remains a question, where does this extra gas arise from?

Surrounding galaxies is a gaseous medium where there is a significant amount of gas in various phases that can play a significant role in galaxy evolution. There are many aspects of this that can and do relate to the evolution of galaxies and to the structural evolution of galaxies. Part of this is that many galaxies have a so-called circumgalactic medium (CGM) which surrounds galaxies out to about 100 kpc (e.g., Tumlinson et al. 2017). The CGM is potentially a significant reservoir of gas that can later cool and form stars within the galaxy itself. Tracing this process is however very difficult, but we can make some progress by investigating the amount of star formation in galaxies as a function of time, as well as the evolution of the gas density and merger history. Thus, in a real sense, by examining the structural evolution of galaxies we can infer the presence of other galaxy formation features, such as gas accretion, and quantify its importance.

One major clue to this from observations of high redshift massive galaxies, with log $M_* > 11\ M_\odot$, is that these galaxies have an average star formation rate which is relatively constant at $1.5 < z < 3$, but declines at $z < 1.5$. Another clue we can use to calculate the amount of material added to galaxies outside of mergers is the fact that the stellar mass almost never decreases with time, such that the value of M_* only goes up.

The increase in stellar mass for a galaxy can occur through the accretion of outside material from mergers ($M_{*,M}(t)$), as we already discussed, but it can also occur through star formation. To calculate the relative amounts we express the increase in stellar mass in a galaxy due to star formation as $\int \psi(t) \times \delta t$, where ψ is the star formation rate within the galaxy as a function of time. Over the redshift range $1.5 < z < 3$, which is about 2 Gyr, the ongoing star formation in a typical galaxy will increase its stellar mass to the point where it approximately doubles, with a typical star formation rate of 100 $M_\odot$ year^{-1}. This high star formation rate means that there has to be a similar amount of mass in cold gas as stars, such that the amount of stars can double by converting this gas into stars. However, the amount of gas mass in these galaxies is only at most 10%–15% on average for typical massive galaxies at this epoch $1.5 < z < 3$. This is a problem because there is not enough gas within the galaxy itself to sustain this star formation over such a long period of time. If we account for the amount of gas brought in through these mergers this is also not enough to add to this gas (e.g., Conselice et al. 2013); something else is needed. This gas must originate from the intergalactic medium or from the CGM surrounding galaxies. We give a calculation below for how this number is computed.

We start this investigation by constructing the continuity equations for tracing the evolution of the cold gas and stellar mass of a galaxy as a function of time. Within this idea we can write the change of stellar mass as a function of time, $M_*(t)$, as

$$M_*(t) = M_*(0) + M_{*,M}(t) + \int \psi \times \delta t - M_{*,\mathrm{recy}}, \tag{5.7}$$

where the value $M_*(0)$ is the initial stellar mass of a typical galaxy under observation at $t = 0$, which usually corresponds to some high redshift; in the example we use this is typically $z \sim 3$. The value $M_{*,\mathrm{recy}}$ is the stellar mass in stars, which is sent back into a gaseous form after stellar evolution. This is produced during different stages of giant stars, planetary nebula, and supernova. Typically, about 15% of this mass is returned back into the ISM of the galaxy. We express how the cold gas mass (M_g), which can be converted into stars, changes in a similar way. The difference here is that the cold gas mass can increase or decrease with time depending on the star formation and gas accretion rate from the IGM or CGM. The continuity equation for this gas mass as a function of time is given by:

$$M_g(t) = M_g(0) + M_{g,M}(t) + M_{g,A}(t) - \int \psi \times \delta t + \varepsilon M_{*,\mathrm{recy}}, \tag{5.8}$$

wherein this equation we have added terms for the amount of cold gas brought in from mergers, $M_{g,M}(t)$ and the amount of cold gas broughs in through accretion, $M_{g,A}$. The amount of stellar mass created in star formation returned to the ISM is again given by ($M_{*,\mathrm{recy}}$). Only a fraction $\varepsilon = 0.001$ of that mass can later be reprocessed into stars, as most of it is shock heated and thus cannot be used again to form stars over the next few Gyr (Bruzual & Charlot 2003).

One of the observational clues we can use to solve these equations is that the gas mass fraction, M_g/M_* is roughly constant over the interval $1.5 < z < 3$ (e.g., Erb et al. 2006; Genzel et al. 2010; Conselice et al. 2013), such that:

$$\frac{M_g(t)}{M_*(t)} \sim \frac{M_g(0)}{M_*(0)}, \tag{5.9}$$

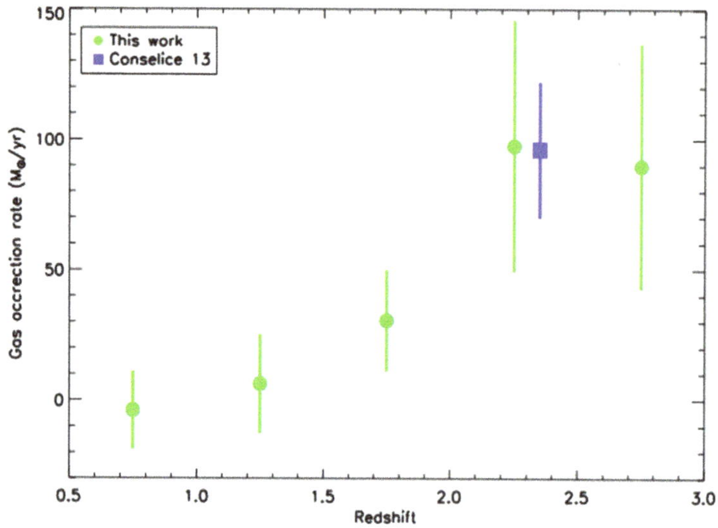

Figure 5.6. Cold gas accretion rate from the intergalactic medium of a constant number density selected sample (Ownsworth et al. 2016). The plotted green circles show the values of the derived accretion with the corresponding error bars denoting the 1σ error on the cold gas accretion based on Monte Carlo methods. The blue square is the average gas accretion rate as found by Conselice et al. (2013) over the redshift range $1.5 < z < 3.0$ using the GOODS NICMOS Survey sample.

Using this observable constraint, we can write the amount of gas mass as a function of time t, $M_g(t)$, by using the above equations. The stellar mass evolution with time can be expressed by replacing $M_*(t)$ into Equation (5.7) to obtain the gas mass as a function of time given by:

$$M_g(t) = M_g(0) \times \left(1 + \frac{M_{*,M}(t)}{M_*(0)} + \frac{\int \psi \times \delta t - M_{*,recy}}{M_*(0)} \right). \tag{5.10}$$

Using this equation, we can calculate that the amount of gas accreted into massive galaxies. The result for massive galaxies is $M_{g,A}/M_*(0) = 0.70 \pm 0.22$ (Conselice et al. 2013) over the entire epoch of $1.5 < z < 3$. Based on this, the amount of gas mass added due to this process is similar to the entire initial stellar mass of a massive galaxy.

This is the gas mass that is added over time, outside of the merger process, in which a galaxy is assembled over the time period $1.5 < z < 3$. This corresponds to a net gas accretion, which is then turned into stars, of $61 \pm 19\, M_\odot\, yr^{-1}$. We must also consider that these galaxies have outflows (e.g., Weiner et al. 2009) that could easily double the amount of gas mass that needs to be accreted from the IGM (e.g., Faucher-Giguère et al. 2011). After considering these outflow the gross inflow rate of gas increases to: $\dot{M}_{acc} = 96 \pm 19\, M_\odot\, yr^{-1}$ (Conselice et al. 2013). Using deeper and wider field data it is now possible to measure this gas accretion rate as a function of redshift, which is shown in Figure 5.6. Quite clearly the gas accretion rate is highest at the highest redshifts and then declines at lower redshifts. By the time we reach $z \sim 1$ we do not find any net gas accretion; clearly this process is limited to the highest redshifts and turns off, or is less important at later times.

These results demonstrate that gas accretion accounts for about half, 49 ± 20%, of the stars which are formed in galaxies at $1.5 < z < 3$. Mergers themselves are responsible for the remainder of the mass assembly, with about half from minor mergers and the other half from major mergers (Bluck et al. 2012; Ownsworth et al. 2014). Gas accretion itself is responsible for 66 ± 20% of all new star formation during this central epoch of star formation, at least for log $M_* > 11$ galaxies. These results and studies demonstrate that gas accretion into massive galaxies at early epochs is potentially a major formation method, which appears to dominate over mergers in terms of importance in the galaxy formation process at $z < 3$.

This empirically measured gas accretion rate is consistent with some theoretical calculations (e.g., Murali et al. 2002; Dekel et al. 2009), as this is predicted to be a major process in galaxy formation. For example, the gas accretion in the model by Murali et al. (2002) predicts a value of $\dot{M}_{g,A} \sim 40\ M_\odot\ yr^{-1}$. Later simulations suggest higher rates close to $\dot{M}_{g,A} \sim 100\ M_\odot\ yr^{-1}$ (e.g., Dekel et al. 2009; Faucher-Giguère et al. 2011). Thus empirical results on measuring cold gas accretion rates are in general agreement with these models.

What the results of this chapter show are that galaxy structure can be used to directly measure processes which are occurring for distant galaxies, namely how much merging is ongoing as a function of time. The merger fraction evolution is confirmed by examining it in several ways including morphology/structure and by looking for galaxies in pairs which will merge within a given timescale. These results then allow us to derive how other modes, which are not readily visible in any observational way, such as gas accretion, are affecting the formation of galaxies through cosmic time. Namely, we are uncovering the build up of stellar mass of galaxies due to different processes. In the next chapter we take a broader view of how galaxy structures can be used to infer formation mechanisms of stars and central black holes of galaxies, as well as for how environment is affecting the ways in which these galaxies are forming.

References

Abraham, R. G., van den Bergh, S., Glazebrook, K., et al. 1996, ApJS, 107, 1
Baldry, I. K., & Glazebrook, K. 2003, ApJ, 593, 258
Bell, E. F., McIntosh, D. H., Katz, N., et al. 2003, ApJS, 149, 289
Bertone, S., & Conselice, C. J. 2009, MNRAS, 396, 2345
Blanton, M. R., Dalcanton, J., Eisenstein, D., et al. 2001, AJ, 121, 2358
Bluck, A. F. L., Conselice, C. J., Buitrago, F., et al. 2012, ApJ, 747, 34
Bluck, A. F. L., Conselice, C. J., Bouwens, R. J., et al. 2009, MNRAS, 394, 51L
Buitrago, F., Trujillo, I., Conselice, C. J., et al. 2008, ApJ, 687, 61L
Buitrago, F., Trujillo, I., Conselice, C. J., et al. 2013, MNRAS, 428, 1460
Brinchmann, J., Abraham, R., Schade, D., et al. 1998, ApJ, 499, 112
Brinchmann, J., & Ellis, R. S. 2000, ApJ, 536, 77L
Bruzual, G., & Charlot, S. 2003, MNRAS, 344, 1000

Carlberg, R. G. 1991, ApJ, 375, 429
Casteels, K. R. V., Conselice, C. J., Bamford, S. P., et al. 2014, MNRAS, 445, 1157
Conselice, C. J. 2009, MNRAS, 399, L16
Conselice, C. J., Bershady, M. A., & Jangren, A. 2000, ApJ, 529, 886
Conselice, C. J., Bershady, M. A., Dickinson, M., & Papovich, C. 2003, AJ, 126, 1183
Conselice, C. J., Blackburne, J., & Papovich, C. 2005, ApJ, 620, 564
Conselice, C. J. 2006a, MNRAS, 373, 1389
Conselice, C. J. 2006b, ApJ, 638, 686
Conselice, C. J., Rajgor, S., & Myers, R. 2008, MNRAS, 386, 909
Conselice, C. J., Yang, C., & Bluck, A. F. L. 2009, MNRAS, 394, 1956
Conselice, C. J., Bluck, A. F. L., Ravindranath, S., et al. 2011a, MNRAS, 417, 2770
Conselice, C. J., Bluck, A. F. L., Buitrago, F., et al. 2011b, MNRAS, 413, 80
Conselice, C. J., Morlock, A., Bluck, A. F. L., Grützbauch, R., & Duncan, K. 2013, MNRAS, 430, 1051
Conselice, C. J. 2014, ARA&A, 52, 291
Conselice, C. J., Twite, J. W., Palamara, D. P., & Hartley, W. 2018, ApJ, 863, 42
Dekel, A., Sari, R., & Ceverino, D. 2009, ApJ, 703, 785
Dickinson, M., Papovich, C., Ferguson, H. C., et al. 2003, ApJ, 587, 25
D'Onghia, E., Mapelli, M., & Moore, B. 2008, MNRAS, 389, 1275
de Ravel, L., Le Fèvre, O., Tresse, L., et al. 2009, A&A, 498, 379
de Vaucouleurs, G., de Vaucouleurs, A., Corwin, H. G., et al. 1991, Third Reference Catalogue of Bright Galaxies (New York: Springer)
Driver, S. P., Windhorst, R. A., Ostrander, E. J., et al. 1995, ApJ, 449, 23L
Duncan, K., Conselice, C. J., Mortlock, A., et al. 2014, MNRAS, 444, 2960
Duncan, K., Conselice, C. J., Mundy, C., et al. 2019, ApJ, 876, 110
Erb, D. K., Steidel, C. C., Shapley, A. E., et al. 2006, ApJ, 646, 107
Faucher-Giguère, C.-A., Kereš, D., & Ma, C.-P. 2011, MNRAS, 417, 2982
Ferreira, L., Conselice, C. J., Duncan, K., et al. 2020, ApJ, 895, 115
Fontana, A., Donnarumma, I., Vanzella, E., et al. 2003, ApJL, 594, L9
Fukugita, M., Hogan, C. J., & Peebles, P. J. E. 1998, ApJ, 503, 518
Genzel, R., Tacconi, L. J., Gracia-Carpio, J., et al. 2010, MNRAS, 407, 2091
Glazebrook, K., Ellis, R., Santiago, B., et al. 1995, MNRAS, 275, L19
Glazebrook, K., Abraham, R. G., McCarthy, P. J., et al. 2004, Natur, 430, 181
Hogg, D. W., Blanton, M., Strateva, I., et al. 2002, AJ, 124, 646
Jeans, J. H. 1919, Problems of Cosmogony and Stellar Dynamics (Cambridge: Cambridge Univ. Press)
Kajisawa, M., & Yamada, T. 2001, PASJ, 53, 833
Kartaltepe, J. S., Sanders, D. B., Scoville, N. Z., et al. 2007, ApJS, 172, 320
Kauffmann, G., Heckman, T. M., White, S. D. M., et al. 2003, MNRAS, 341, 54
Kelvin, L. S., Driver, S. P., Robotham, A. S. G., et al. 2014, MNRAS, 444, 1647
Kennicutt, R. C. 1989, ApJ, 344, 685
Kinney, A. L., Calzetti, D., Bohlin, R. C., et al. 1996, ApJ, 467, 38
Le Févre, O., Abraham, R., Lilly, S. J., et al. 2000, MNRAS, 311, 565
Lilly, S. J., Tresse, L., Hammer, F., Crampton, D., & Le Fevre, O. 1995, ApJ, 455, 108L
Lotz, J. M., Davis, M., Faber, S. M., et al. 2008, ApJ, 672, 177L
Lotz, J. M., Jonsson, P., Cox, T. J., & Primack, J. R. 2010, MNRAS, 404, 590

Madau, P., Ferguson, H. C., Dickinson, M., et al. 1996, MNRAS, 283, 1388
Madau, P., & Dickinson, M. 2014, ARA&A, 52, 415
Man, A. W. S., Zirm, A. W., & Toft, S. 2016, ApJ, 830, 89
Mantha, K. B., McIntosh, D. H., Brennan, R., et al. 2018, MNRAS, 475, 1549
Marzke, R. O., Geller, M. J., Huchra, J. P., et al. 1994, AJ, 108, 437
Mortlock, A., Conselice, C. J., Hartley, W., et al. 2013, MNRAS, 433, 1185
Mortlock, A., Conselice, C. J., Hartley, W. G., et al. 2015, MNRAS, 447, 2
Mundy, C. J., Conselice, C. J., Duncan, K. J., et al. 2017, MNRAS, 470, 3507
Murali, C., Katz, N., Hernquist, L., et al. 2002, ApJ, 571, 1
Ownsworth, J. R., Conselice, C. J., Mortlock, A., et al. 2014, MNRAS, 445, 2198
Ownsworth, J. R., Conselice, C. J., Mundy, C. J., et al. 2016, MNRAS, 461, 1112
Patton, D. R., Pritchet, C. J., Yee, H. K. C., Ellingson, E., & Carlberg, R. G. 1997, ApJ, 475, 29
Patton, D. R., Pritchet, C. J., Carlberg, R. G., et al. 2002, ApJ, 565, 208
Spergel, D. N., Verde, L., Peiris, H. V., et al. 2003, ApJS, 148, 175
Steidel, C. C., Adelberger, K. L., Giavalisco, M., et al. 1999, ApJ, 519, 1
Trujillo, I., Conselice, C. J., Bundy, K., et al. 2007, MNRAS, 382, 109
Tumlinson, J., Peeples, M. S., & Werk, J. K. 2017, ARA&A, 55, 389
van den Bergh, S., Cohen, J. G., & Crabbe, C. 2001, AJ, 122, 611
Weiner, B. J., Coil, A. L., Prochaska, J. X., et al. 2009, ApJ, 692, 187
Weinzirl, T., Jogee, S., Conselice, C. J., et al. 2011, ApJ, 743, 87
Whitney, A., Conselice, C. J., Bhatawdekar, R., et al. 2019, ApJ, 887, 113
Windhorst, R. A., Taylor, V. A., Jansen, R. A., et al. 2002, ApJS, 143, 113

Christopher J Conselice

Chapter 6

Secondary Products of Evolution

This chapter discusses how the merger history as measured through structures and pairs relates to other aspects of galaxy assembly, such as morphological evolution, the evolution of the star formation rate, and the formation and growth of massive black holes in the centers of galaxies through cosmic time. This chapter also includes a discussion of how other methods such as environment and gas accretion are affecting galaxy evolution and formation along with merging. Overall, this chapter is focused on secondary products of structural evolution, namely physical processes related to how galaxy structure changes with time.

6.1 Interaction and Merger Induced Star Formation

Based on the previous chapter we have a good idea of the galaxy merger rate, and the number of mergers galaxies are involved with as a function of time. We have also investigated how much stellar mass is added to galaxies in the galaxy formation process. This calculation also gives us some idea about the amount of gas mass that must arise from the intergalactic medium.

Using this information we can now ask more detailed questions, including: how much mass is added to galaxies through star formation induced in the merger process? This question of course has some sub-questions including how much of this star formation is added due to major mergers as opposed to minor mergers. Theory predicts that during the merger process star formation rates should change over the merger and at some points have very high values (Figure 6.1). Another question is how much do these mergers add to the formation of central black hole masses and the triggering of active galactic nuclei (AGN)? This is an area where significant research will still need to occur, as we only have some vague ideas about these questions and often conflicting results.

Star formation rates within galaxies are typically measured by using light emitted from galaxies at certain wavelengths; typically and most commonly, ultraviolet or the far-infrared, which directly trace star formation rates by examining young

doi:10.1088/2514-3433/abb602ch6

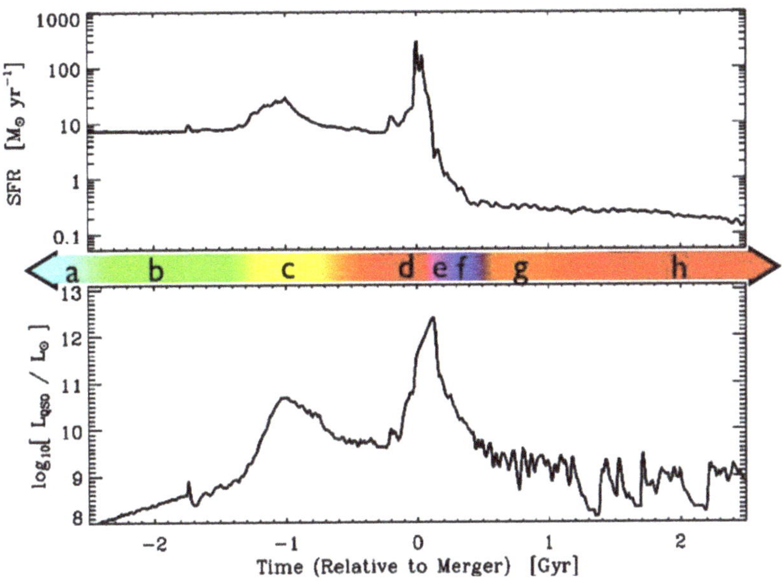

Figure 6.1. Plot reproduced from Hopkins et al. (2008) showing how in theoretical models of merging galaxies that the star formation rate (top panel) varies during the merger process and is only high at certain times. It is therefore possible to locate mergers that do not have a high star formation rate as they are not being observed within this relatively narrow time-span of the merger when the star formation rate is high.

massive stars or dust emission (Kennicutt et al. 1994). It is fairly easy to show that we can identify most of the stellar mass forming in the universe through tracing the star formation rate, which can be probed up to redshifts of $z = 7$ (e.g., Madau & Dickinson 2014). However, it is important to recognize a few things about these measured star formation rates.

One of these is that all stars have to be created in star formation events; without star formation there is no creation of new stars. Galaxies cannot exist without this star formation, which makes up the mass that we can observe from galaxies. It is important to realize that just because star formation happens in an individual galaxy that does not necessarily mean those stars remain in that galaxy as an isolated system from the start of its evolution until now. Galaxy interactions and mergers refer to the formation of matter through the merger process, whereby the assembly of galaxies can occur after star formation by adding two separate galaxies together to form a new system. At the same time, it is possible or even likely that galaxy mergers and interactions will induced star formation in galaxies. The reason for this is simple—as two galaxies interact or merge, the gas clouds within those galaxies will be induced to collide with each other and become denser, which can then form into stars. This is a natural outcome of galaxies which are interacting and merging.

It is not well understood, or generally agreed upon, what the role of interactions and mergers are in triggering star formation in galaxies. Galaxy mergers and interactions are often observed in nearby groups, which may be a common way to induce star formation in these systems. Thus the question is—do galaxy interactions and mergers trigger star formation in galaxies? This can be studied through data both in the local universe, and also at higher redshifts. When examining interacting galaxies the star formation rate for pairs separated by a few 10s kpc is ~80% higher than in galaxies that are not interacting (Robaina et al. 2009). However, starbursting galaxies, where a real excess of star formation is found over normal galaxies, are often correlated with merging and interaction events. This evidence shows that the most intense star formation is often triggered by mergers. This is a recurring theme which we also find with AGN, such that mergers do not seem to generally increase on average the star formation rate by a large amount. However, the most star-forming galaxies, or extreme examples, are indeed those undergoing mergers. It is also the case that obscured Type-II Compton-thick AGN are more likely to show evidence for merging activity (e.g., Kocevski et al. 2015). Therefore, just like star formation and merging activity, the correlation with AGN and mergers may only occur during limited parts of the merger phase.

Observations show that the average enhancement of star formation in merging galaxies is modest, with most of the star formation occurring at $z < 3$ arising from gas brought in through accretion rather than merging activity (e.g., Knapen et al. 2015; Lofthouse et al. 2017). It is however important to realize that the triggering of star formation is only one way in which mergers are influencing the formation of galaxies. The process of mass assembly of already existing galaxies which merge together is very likely the dominant way in which mergers influence galaxy formation. As we discussed in Chapter 5 the prevalence of mergers over star formation occurs at $z > 5$, and this is where the location of intense starbursts triggered by mergers is likely to be found.

6.2 Interaction Induced Central Black Hole Formation

We can also investigate if and how the formation and fueling of AGN are triggered by merging. The idea is relatively simple—when we have two galaxies merge this presents an unstable situation in which gas can flow through a galaxy. Some of that gas could very easily be collated or collected toward the center. When this occurs some of that gas will enter the accretion disk of an AGN and thereby later fuel it. The effect of this would be the production of active galaxies such as quasars and other types of AGN. This ultimately will build up the mass of the central massive black holes in these systems.

Whether or not AGN are triggered by mergers and interactions is a long-standing question with a mix of answers depending on what type of source is looked at and how luminous it is. It has been known for sometime that quasars, which are the most luminous AGN, are often in the centers of galaxies that have recently undergone mergers (Bahcall et al. 1997; Urrutia et al. 2008). This is also seen in deep ALMA imaging of extremely high redshift sources at $z > 7$ (Bañados et al. 2019). There is no universal agreement on this observation however with others finding that dusty Type II quasars have a merger fraction of ~25% (Schawinski et al. 2012). This is a higher merger fraction than what is found for field galaxies, but it certainly is not a large fraction of all of these quasars. This is also the case for typical X-ray selected AGN, as seen in deep HST field imaging, such that only a small fraction of these systems are undergoing mergers (e.g., Kocevski et al. 2012).

However, just as we saw for star formation, it is likely that a large part of these conflicting results on the merger fraction for AGN depends on the luminosity of the sources and the underlying masses of the host galaxies. The most luminous sources are inevitably often mergers, while lower luminous sources can be triggered in many other ways including through gas accretion. Thus in general it appears to be a fundamental truth that activity is produced by mergers and interactions in the most luminous and intense systems, but they are less important for more common systems.

6.3 Environmental Effects on Structure

It is now well known that galaxy morphology and structure correlate strongly in the local universe with a galaxy's environment (e.g., Dressler 1984; Postman et al. 2005). In general, early-type and elliptical galaxies are found in the central regions of clusters (Figure 6.2). It is therefore worth investigating how this fits in with ideas about galaxy structural evolution. One of the major correlations between galaxy environment and other physical properties of galaxies is with the local (as measured

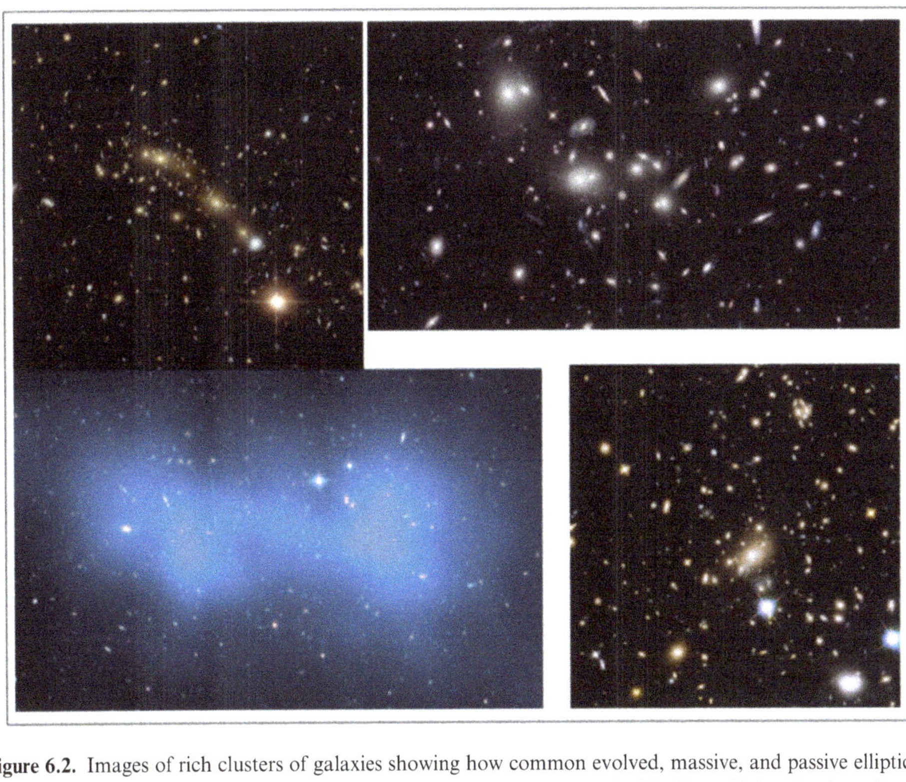

Figure 6.2. Images of rich clusters of galaxies showing how common evolved, massive, and passive ellipticals are in the centers of these systems. The result of this is a systematic trend of finding a higher fraction of early-type galaxies in denser environments, and spiral galaxies in lower density environments. Images courtesy of NASA/ESA.

in galaxies per projected Mpc^2) and perhaps even the global environment in which a galaxy is located. Note that this does not mean that the structure of a galaxy is determined by the environment in which it forms or evolves within, but by the environment in which we find it when we observe it. Mostly these observations arise in the nearby universe where we can study in great detail how the structure and morphology of a galaxy relate distinctly to its environment.

Historically, it was noticed by early practitioners of galaxy studies that early-type galaxies, or ellipticals, were much more likely to be found within dense areas or denser collections of galaxies. At that time it was not clear that clusters existed, but even Messier and the Herschels' early catalogs of nebulae showed that systems were clustered in the sky. This suggests that they are related and correlate with each other in some way. Today we know that environment is a fundamental aspect for understanding how galaxies form and evolve.

In terms of modern work, it was the paper Dressler (1980) that really set the stage for the study of how galaxy environment and morphologies are related to each other. What Dressler (1980) found was that within denser *local* environments (denoted by Σ) galaxies were more likely to have an early-type or elliptical morphology (see Figure 6.3 for a modern version of this). This relation is such that within more dense local

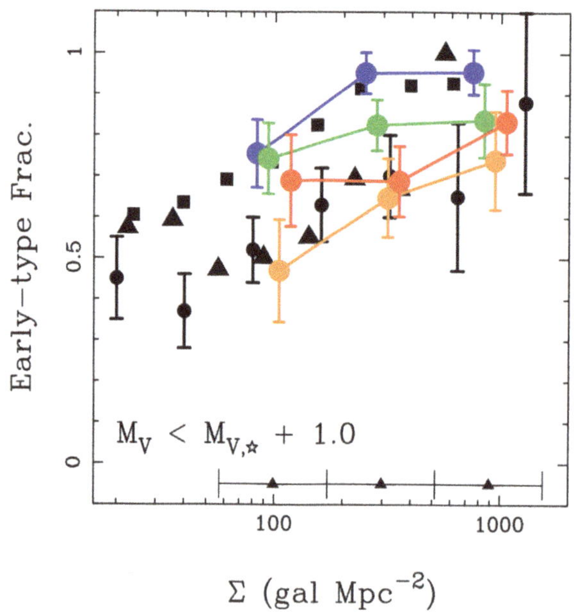

Figure 6.3. The morphology-density relation as parameterized by the galaxy local density parameter (Σ) as a function of the fraction of galaxies in a given environment which are early-type systems (reproduced from Holden et al. 2007). The galaxies plotted on here are those one magnitude brighter than M_V^*—the characteristic luminosity. The clusters within this sample are Coma at $z = 0.023$ (blue circles; from classifications by Dressler 1980), CL 1358 at $z = 0.328$ (green circles), MS 2053 at $z = 0.587$ (orange circles), and a combination of MS 1054 and RX J0152 (at $z = 0.831$ and $z = 0.834$) (red circles). The results from Dressler et al. (1997) for galaxies at $z \sim 0.4$ are shown as black triangles, and Dressler (1980) for nearby systems as black solid boxes.

environments, the more likely a galaxy in that environment will be early-type and non-star-forming (e.g., Dressler 1984; Gomez et al. 2003; Blanton & Moustakas 2009). Disk galaxy properties also correlate with environment such that few classical bulge or elliptical systems are found in low density environments (Kormendy et al. 2010). It is important to keep in mind that these correlations are for the local environment of a galaxy, not its global environment. In this way, it is possible to find spiral galaxies in rich clusters of galaxies, but for the most part only in the outer portions of these clusters where the local environmental density is low.

Since these studies were first published it has been debated what is the natural origin of this relation—is it nature or nurture? That is, are we seeing this relationship because galaxies in dense environments evolve to become ellipticals from some ancestral galaxies, or do galaxies in dense environments just naturally form quickly into ellipticals because they are in a dense environment? This is similar to the debate about the origin of galaxies in general, whereby we ask whether a galaxy forms late due to merging, or early due to a rapid collapse.

We can try to understand this issue by examining the basic idea for how galaxies form in the context of different environments and densities of galaxies. The essential idea, expressed in more detail in early chapters, is that galaxies assemble early in the history of the universe once matter collects into gravitationally bound systems. Most of this matter is in the form of gas and dark matter. We can predict how this process of galaxy formation occurs through the dominant ideas of cosmology using the Λ-Cold Dark Matter (ΛCDM) cosmology. The Λ term describes the existence of dark energy and the cold dark matter gives the temperature of the dark matter. We have some ideas of the relative abundance of dark matter and dark energy based on recent measurements, such as the Planck cosmic background satellite (Planck Collaboration et al. 2014).

In the ΛCDM cosmology the first galaxies are small low mass ones due to the fact that the dark matter temperature is "cold." This is due to the fact that gravity cannot act over large distances before the first objects form, therefore they tend to be lower mass systems. Within this paradigm, the larger objects form from the mergers of these smaller objects, which we described as the hierarchical assembly of matter. This is also the way in which galaxy clusters are formed by the accumulation of galaxies through gravity.

A majority of galaxies in the nearby universe are in groups, therefore galaxy evolution and formation do not occur in isolation except for a very small fraction of systems. What is also found is that galaxies at all redshifts up to $z \sim 6$ cluster on the scale of nearby groups, at the level of $\sim$1 Mpc (Gerke et al. 2005). Using simple dynamical friction arguments it is easy to show that galaxy mergers should be relatively common in these environments, and a dominant process for driving evolution.

In general, galaxy evolution occurs via both internal and external drivers. The environments in which a galaxy finds itself must often have a strong effect on how some galaxies form and evolve. An example of this is those galaxies surrounded by other galaxies in close proximity (see Figure 3.5). In this case gravitational interactions and mergers can easily induce star formation and dynamical processes can remove stars/gas through tides. As such, the masses of galaxies can increase through the accretion of satellites and the number of galaxies decreases due to mergers. Furthermore, there can be changes to the structures and morphologies of galaxies due to this close proximity. Galaxy groups also often contain an intragroup medium—gas in the space between the galaxies in a group. Because of this, the intragroup gas can interact with gas within galaxies themselves which can then be removed through processes such as ram-pressure stripping and exhaustion. When this happens, and because gas cannot be replenished after all the existing gas is used up in star formation, a galaxy will become "passive" and lose its structure due to faded star formation.

We can scale the relative importance and strength of these various processes by the distribution of galaxy velocities within a group or cluster. This is denoted often by the velocity dispersion of its members, σ, which gives a quantitative measurement of the range of velocities of individual galaxies within a group. The removal of gas in a galaxy through interacting with the intragroup medium is called ram-pressure stripping. The resulting force from this stripping is given by $\sim\rho\sigma^2$, where ρ is the density of the intragroup medium's gas. Since the velocity dispersions of groups are typically $\sim$250–400 km s^{-1}, the efficiency of this process in groups is not large compared with clusters of galaxies where we find much higher values of both ρ and σ. There are also high-speed galaxy interactions between galaxies in a dense environment in a process called galaxy harassment (e.g., Moore et al. 1996). This process is unlikely to produce significant effects in groups due to the relatively low velocities between member galaxies; this process is most efficient in rich clusters of galaxies where the velocity dispersion is large. Thus properties of environment are more likely to have an effect on a galaxy's structure in denser clusters than in groups.

Another possible environmentally driven process is galaxy exhaustion, whereby a galaxy uses all of its gas to form stars but that gas is not replenished through for example mergers with low mass galaxies or gas accretion described in Chapter 5. This is however not a fast process, with a timescale of about a $\sim$ Gyr in clusters. This process also requires a removal mechanism, such as the intracluster medium, that stops any further gas accretion onto galaxies. This process is potentially very important in clusters where the density of the intracluster gas is high. These mechanisms are however not as effective in groups, observationally we can see this as we observe star formation in member galaxies.

Galaxy mergers should be relatively common in groups of galaxies compared to rich clusters. This is due to the low-velocity dispersion of galaxies in groups and the close proximity of members. Different types of nearby groups, such as compact groups and fossil groups, are likely the result of the galaxy merger process. This can be seen by considering dynamical friction calculations which demonstrate that the merger timescale varies as $\tau_{\rm merge} \sim \sigma^3$. Thus galaxies within groups with lower velocity dispersions will merge quicker than galaxies in clusters. Groups with velocity dispersions >300 km s^{-1}, the limit from which anything higher would be considered galaxy clusters, will not undergo a significant number of mergers over a Hubble time. This is the primary reason why galaxy mergers must occur in groups of galaxies for the most part.

Detailed simulations and analytical calculations show that the dynamical friction process in groups changes these systems such that they become more centrally concentrated by about a factor of two since $z \sim 1$. This effect is strongest for groups with low velocity dispersions, <150 km s^{-1}. Merging in groups at $z \sim 0.4$ should be low, about 2% per Gyr per group.

Within groups of galaxies, which constitute the environments for most galaxies, the merger rate should increase with look back time at $z > 1.5$ when the universe's density was much higher. This is due to the fact that the universe was much denser and galaxies were physically closer together in this epoch, which naturally causes more mergers to occur. When we consider a Λ-dominated universe, the mass density increases as the Hubble parameter squared, or $H^2 \sim (1 + z)^3$, with a corresponding merger rate which can be expressed as $\sim\sqrt{(H^2)} \sim (1 + z)^{1.5}$. See Figure 6.4 for a quick summary of these facts for groups of galaxies.

The evolution of this rate of mergers can be used to explain the range of galaxy types in groups, such as the mixture of early-types and spirals we observationally see. Ellipticals and bulges in these early-type systems were formed through mergers early in the universe, as outlined in Chapter 5. Later on, as evolution continued to occur through for example gas accretion, then disk+bulge systems formed. Thus, from this it appears that the way in which galaxies in dense environments form is through the merger process for the most part. We also find that more massive galaxies are more likely to be early-type such that there exists a strong relationship between morphology and stellar mass, whereby the most massive galaxies tend to have elliptical morphologies with little ongoing star formation. When we combine this fact with the morphology-redshift relation it shows that the structure of a galaxy depends upon its mass, local environment, as well as time. Peng et al. (2010) demonstrated this by showing that the color and star formation rate for galaxies depends both upon their environment, as well as their stellar mass (see Figure 6.5).

Galaxy Mergers Notes

- Mergers should be common in groups - dynamical friction time-scale evolves as (group σ^3)
- ~250 km/s upper limit for dragging galaxies into the centers of groups over a Hubble time
- Low redshift merger rate expected to be low, around 2% of galaxies in groups merge per Giga-year
- At higher-redshift the mass density increases as $H^2 \sim (1+z)^3$. Resulting in a higher merger rate of $\sqrt{(\rho)} \sim (1+z)^{1.5}$

Figure 6.4. The summary of properties of galaxies in groups and how these relate to the likely properties of mergers. Because of these features, galaxy mergers are more likely to occur in the environment of groups of galaxies than in any other location or environment in the universe.

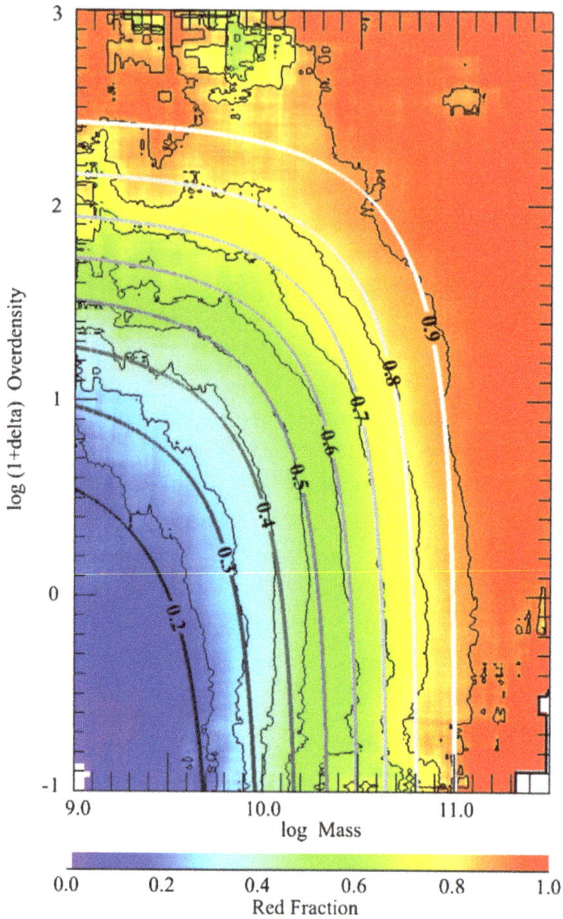

Figure 6.5. Peng et al. (2010) demonstrating how galaxy colors, which are a reflection of morphology and structure, correlate with both stellar mass and local density, which in this case is denoted by (1 + delta). Bluer colors show where more star-forming galaxies are located and the redder colors show the location where more passive/red galaxies are located in this space. As can be seen, both high mass galaxies and those in dense environments (higher delta values) produce galaxies that are more passive/red. Reproduced from Peng et al. (2010).

As can be seen in this figure, both a high stellar mass and a high density can produce passive/red galaxies.

The question is which relationship—mass or environment—is more fundamental (e.g., Grützbauch et al. 2011a, 2011b)? This relates to, and is another way of addressing, the nature versus nurture problem for galaxy formation. We can address this problem by examining the structures and morphologies of galaxies at higher redshift overdensities, or within proto-clusters that are just forming (e.g., Dressler et al. 1997; Holden et al. 2007; Sazonova et al. 2020). However, observations of overdensities at high redshifts $z > 2$ are just starting in earnest, but already the observations give us some ideas of how this evolves. What we know so far is that massive clusters at high redshifts, up to $z \sim 1.2$, display a similar morphology-density relation as local galaxies, such that denser areas of galaxies contain higher fractions of elliptical systems. However, what is also seen is that the normalization is lower, such that at a given density, Σ, there are fewer ellipticals than at the same local density at lower redshifts (Figure 6.3).

What we would like to also know is how this evolution with environment occurs for more typical galaxies found in the "field"—that is not within a group or cluster environment. For these non-cluster galaxies, which includes galaxy groups, Tasca et al. (2009) examine the morphology-density relationship for 100,000 galaxies in the COSMOS ACS HST survey field. Tasca et al. (2009) showed that the morphology-density relation evolves slightly with redshift, becoming less strong at higher-z (see also Grützbauch et al. 2011a, 2011b). What they find is that above a stellar mass of $10^{10.6}$ $M_{\odot}$ the structures of galaxies are more determined by the stellar mass as the critical factor rather than density (e.g., Grützbauch et al. 2011a). For rich clusters at $z \sim 1$, and the formation of S0s within them, the observed trend of structure with environment is not as steep as is found at $z \sim 0$. This suggests that S0s are not entirely formed yet in these clusters, however the elliptical population does seem to be in place compared with the population at $z \sim 0$, suggesting that for at least the massive ellipticals their formation is driven by "Nature."

We can go beyond simple morphology and examine the structural aspects of galaxies, including the evolution of galaxy size and how it varies with environment. The morphology results are also backed up by finding a kinematic-structure/environment relation (Brough et al. 2017). There are still a limited number of investigations of size evolution with environment, especially the quantification of this with structural parameters. These studies typically find that galaxies in higher density environments at $z > 0.5$ show signs of a more rapid increase of galaxy size with redshift in comparison to the field (e.g., Lani et al. 2013). This is also seen in clusters at $z > 1$ where the morphology-density relation is not yet settled. In clusters where this relation is not seen, merging galaxies are much more common (e.g., Sazonova et al. 2020)—another indication for how environment drives the evolution of galaxies at these early times. This is an indication that a dense environment can drive a more rapid galaxy evolution, although this is still only measured to be a slight effect that needs further confirmation and quantification.

6.4 Summary

In this chapter we discuss the "value-added" properties of mergers—that is, how mergers can induce star formation as well as AGN activity and supermassive black hole growth in galaxies. The observations of this are ambiguous still in the sense that there are conflicting results in the literature. However, what does appear to be the case is that the most massive systems with the most intense star formation and AGN activity (such as in QSOs) are more likely to be involved in merging activity.

We also discuss the role of environment in establishing the morphologies and structures of galaxies. This relates in many ways to the morphology-density relationship such as that in dense environments we are more likely to find elliptical galaxies, and spiral galaxies are more likely to be found in lower-density environments. This suggests that environment either induces structural changes in galaxies or morphologies are established because galaxies form early in a given environment. This chapter also shows how the environment is important for inducing merging activity in galaxies such that the group environment is the ideal one for mergers to occur in, while clusters have too high a velocity dispersion for mergers to be effective.

References

Bahcall, J. N., Kirhakos, S., Saxe, D. H., et al. 1997, ApJ, 479, 642
Bañados, E., Novak, M., Neeleman, M., et al. 2019, ApJL, 881, L23
Blanton, M. R., & Moustakas, J. 2009, ARA&A, 47, 159
Brough, S., van de Sande, J., Owers, M. S., et al. 2017, ApJ, 844, 59
Dressler, A. 1980, ApJ, 236, 351
Dressler, A. 1984, ARA&A, 22, 185
Dressler, A., Oemler, A., Couch, W. J., et al. 1997, ApJ, 490, 577
Gerke, B. F., Newman, J. A., Davis, M., et al. 2005, ApJ, 625, 6
Gomez, P. L., Nichol, R. C., Miller, C. J., et al. 2003, ApJ, 584, 210
Grützbauch, R., Conselice, C. J., Varela, J., et al. 2011a, MNRAS, 411, 929
Grützbauch, R., Chuter, R. W., Conselice, C. J., et al. 2011b, MNRAS, 412, 2361
Holden, B. P., Illingworth, G. D., Franx, M., et al. 2007, ApJ, 670, 190
Hopkins, P. F., Hernquist, L., Cox, T. J., et al. 2008, ApJS, 175, 356
Kennicutt, R. C., Tamblyn, P., & Congdon, C. E. 1994, ApJ, 435, 22
Knapen, J. H., Cisternas, M., & Querejeta, M. 2015, MNRAS, 454, 1742
Kocevski, D. D., Faber, S. M., Mozena, M., et al. 2012, ApJ, 744, 148
Kocevski, D. D., Brightman, M., Nandra, K., et al. 2015, ApJ, 814, 104
Kormendy, J., Drory, N., Bender, R., & Cornell, M. E. 2010, ApJ, 723, 54
Lani, C., Almaini, O., Hartley, W. G., et al. 2013, MNRAS, 435, 207
Lofthouse, E. K., Kaviraj, S., Conselice, C. J., et al. 2017, MNRAS, 465, 2895
Madau, P., & Dickinson, M. 2014, ARA&A, 52, 415
Moore, B., Katz, N., Lake, G., et al. 1996, Natur, 379, 613
Peng, Y.-j., Lilly, S. J., Kovač, K., et al. 2010, ApJ, 721, 193
Planck Collaboration, Ade, P. A. R., Aghanim, N., et al. 2014, A&A, 571, A16
Postman, M., Franx, M., Cross, N. J. G., et al. 2005, ApJ, 623, 721
Robaina, A. R., Bell, E. F., Skelton, R. E., et al. 2009, ApJ, 704, 324
Sazonova, E., Alatalo, K., Lotz, J., et al. 2020, ApJ, 899, 85
Schawinski, K., Simmons, B. D., Urry, C. M., et al. 2012, MNRAS, 425, L61
Tasca, L. A. M., Kneib, J.-P., Iovino, A., et al. 2009, A&A, 503, 379
Urrutia, T., Lacy, M., & Becker, R. H. 2008, ApJ, 674, 80

AAS | IOP Astronomy

The Cosmic Evolution of Galaxy Structure

Christopher J Conselice

Chapter 7

Applications of Structural Evolution

As we discussed in previous chapters, galaxy evolution processes can be seen in the structures of galaxies. As discussed in-depth, we can locate and measure the merger history with structure. Galaxy structure can also reveal other formation mechanisms such as gas accretion, especially at early times, which is inferred by the merger and star formation processes that we can measure partially through using galaxy structure. It turns out that these properties also relate to the underlying cosmology of the universe as a whole. The stellar mass distribution of galaxies and how these evolve also relate to the cosmological properties of the universe. Investigating these features of the universe itself, and using galaxy structural evolution as a tool for learning about these problems is still in its infancy, but it will grow with new data to become a major way to investigate these questions in the future with upcoming large surveys such as with the James Webb Space Telescope, Euclid, Rubin, and Roman telescopes.

7.1 Cosmological Parameters and Galaxy Structure

The dominant model for galaxy formation, as well as the cosmology of the universe, is one in which the universe is hierarchical, as we have described in previous chapters in this book. In fact, as we have shown, the observations of galaxy structural evolution back up this idea of a hierarchical universe. Galaxies and cosmology are in fact intrinsically related, and one affects the other in our universe. To fully appreciate the fact that galaxies are the constituents within the universe, we have to consider them in a cosmological context, which is often done in simulations. However, we can flip this around such that the evolution and formation of galaxies can be used to determine features of the underlying cosmology and galaxy evolution process.

This is such that the structure and formation of galaxies strongly depends on cosmological parameters, including properties of the dark matter. It turns out that different cosmological parameters and different universe models produce different

doi:10.1088/2514-3433/abb602ch7

types of galaxies, as well as when and how they form. It is thus possible to use observables such as the galaxy structure-redshift relation as a method for determining the cosmology of the universe itself.

The idea behind this is fairly simple and can be explained in a few examples. The universe and thus also the galaxies within it have an evolution which is dominated by the cosmological features of the matter density Ω_m and the density of dark energy which is parameterized by the parameter Ω_Λ. If the universe is dominated by dark energy early in the history of the universe it would be accelerating faster and the expansion of space would have happened sooner than what has appeared to have happened within our own universe (Turner & Riess 2002). This would naturally lead to a universe where galaxies become more separate from each other earlier, driving down the merger rate at later times (e.g., Conselice et al. 2014). Thus, a universe dominated by dark energy would have less mergers than one dominated by dark matter.

The essential feature here is that the dark energy and dark matter compete to drive the regulation for how much and when galaxies in the universe are forming and evolving. A universe that is expanding and accelerating will produce systems that are further apart, and if this expansion is very rapid in a short period of time, relatively speaking, then galaxies will not evolve or form as much due to the fact that they are further apart from each other. On the other hand, if the universe was heavily dominated by matter with very little dark energy, then the expansion would be slowed down, galaxies would become closer together, and more galaxy formation and merging would occur over time.

Based on this basic outline, it is clear that galaxy mergers can play a key role in testing these ideas, and also refining our knowledge of cosmology. Thus, galaxy mergers are not just important for producing galaxy evolution, and triggering a significant star formation and AGN activity, but also their history and the history of galaxy structure can be used as a measurement of cosmological parameters.

Overall, we know from studies such as Planck and WMAP that the following condition holds:

$$\Omega_m + \Omega_\Lambda = 1. \tag{7.1}$$

A basic kinematic analysis shows that the acceleration of the universe must have started at redshifts of $z \sim 0.7$ (Turner & Riess 2002). This is due to the relative contribution of matter and dark energy. If the universe was more dark matter dominated and had no dark energy then the effect will be such that mergers will be much more common, and galaxies today would be much more massive than they actually are and would have experienced more mergers. In fact, finding high redshift galaxy mergers and tracing their evolution was originally instigated to find a way to quantify these parameters (Carlberg 1991).

Thus, by comparing the contribution of dark matter and dark energy to the universe's energy budget one gets a unique value of the merger history (see Figures 7.1 and 7.2). This was examined in detail in Conselice et al. (2014) who showed using dark matter halo merger histories within different cosmological

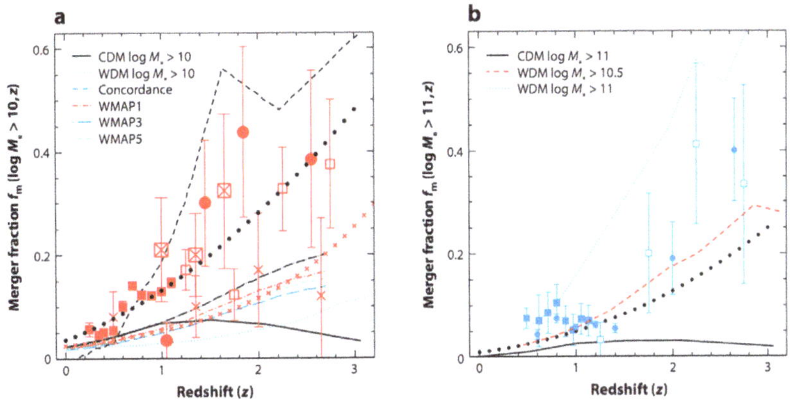

Figure 7.1. A demonstration of the use of the merger fraction evolution as a probe of dark matter particle temperature. Here we have plotted the merger fraction as a function of redshift up to $z \sim 3$ for galaxies with stellar masses (a) $M_* > 10^{10}$ $M_\odot$ and (b) $>10^{11}$ $M_\odot$. Also over plotted are the predicted merger fractions for dark matter halos in different cosmologies, including the WMAP1,2,3 year results, the concordance cosmology with $\Omega_m = 0.3$, $\Omega_\Lambda = 0.7$ and $H_0 = 70$ km s^{-1} Mpc^{-1}. As can be seen from this diagram, the warm dark matter models give a better fit than this particular CDM model.

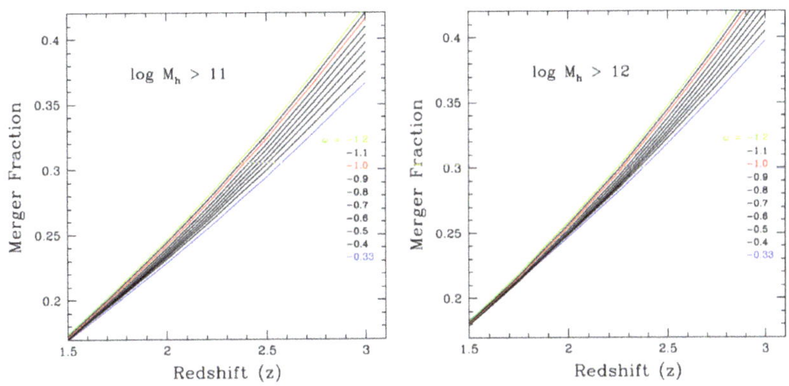

Figure 7.2. Model merger fractions derived for different cosmologies using the parameterization of the dark energy equation of state, as given by ω (see text). As can be seen, the values of the merger fractions do not differ very much between the different ω values, and a much better accuracy on merger history uncertainties will be required to distinguish between likely values of ω.

models produces different predicted merger histories. Conselice et al. (2014) found that variations on the relative amounts of dark energy and dark matter produce different merger histories. While it is possible to rule out certain cosmologies with this approach, such as the Einstein–de Sitter universe, there are very small differences in predicted merger fractions for realistic possible cosmologies. This will however be a possible test to make for a range of realistic cosmologies once merger fraction errors shrink after large area surveys are carried out by the Euclid Space Telescope and Rubin to measure merger histories with very good statistics. The available merger data is not currently competitive with measurements from standard methods using for example the cosmic background radiation, giving errors on the cosmological parameters which are a tenth of the errors from using the merger fraction. This is partially due to limited area surveys that can currently be used for this type of analysis, but systematics in measuring mergers will have to be improved as well. The best fit value using cosmology currently gives: $\Omega_\Lambda = 0.84^{+0.16}_{-0.17}$ (Conselice et al. 2014).

Doing better than this value will require an order of magnitude improvement in surveys for mergers than what we can do now. The dark energy evolution is often parameterized as an equation of state given by: $\omega = P/\rho$. It is well known that ω is ~ -1 and any deviation from -1 would have to be very small. As Figure 7.2 shows, to obtain reasonable differences between $\omega = -1$ and $\omega = -0.9$ requires the errors on the merger fractions to be around $\delta A \sim 0.01$. This is possible if we can measure accurate errors on merger fractions and also have enough galaxies in the sample to remove significant statistical fluctuations or random errors.

Upcoming large area imaging and spectroscopic surveys such as Euclid, Rubin, 4MOST, and Roman will possibly allow us to do this test. Whether or not this will become a competitive measurement of cosmology that can measure cosmological properties and parameters at a similar level using galaxy clustering, weak lensing, or the cosmic background radiation remains to be seen, but it seems possible. Why should we even attempt this if it cannot compete? Using mergers to measure cosmology is an alternative and critically independent way of measuring these features that do not depend on any properties of the physics, or the assumptions, which dominate these other methodologies. This method is unique and complementary to these methods, and thus it could be used as a new test on our ideas of cosmology and consistency with galaxy formation itself.

7.2 The Dark Matter Particle Temperature

Another way in which we can study features of the universe with galaxy structures and mergers is through measuring the dark matter particle mass and temperature. As described earlier, within the ΛCDM universe galaxy mergers are common due to there being many small mass galaxies and halos within this cosmology. This is even more common in the early universe before hierarchical structure formation has occurred.

On the other hand, if the dark matter particle was not 'cold' but was 'warm' or even 'hot' we would see a vastly different merger history for galaxies and thus also in the structural evolution of galaxies. Conselice et al. (2014) demonstrated how the merger history can change for merger fractions within different universes made up of Warm Dark Matter (WDM) as compared to Cold Dark Matter (CDM). Within WDM the power spectrum of matter is not as steep at the low mass end compared to CDM. This results in relatively fewer low-mass galaxies per giant galaxy. Because there are fewer low mass galaxies then the merger fraction is not as high as what is obtained with CDM. Of course, at the time of writing, the dark matter particle identity is unknown, and thus it is impossible to use this information to know which model is correct, and different observational evidence has pointed toward and against WDM as an alternative to CDM. This remains an issue that is yet to be resolved.

Can we thus use the merger history as another way to determine the possible temperature and thus the mass of the dark matter particle? Figure 7.1 shows the merger history as published in Conselice et al. (2014) based on morphological analyses overlaid with models of merger fraction histories using CDM and WDM models (see Figure 7.1). Shown on these figures as the dotted blue line is the predictions for major merger fractions for WDM models (e.g., Menci et al. 2012), which do a better job than CDM in matching the data. However, CDM models are a better fit when minor mergers are taken into account. As far as we know from simulations, the observed merger fraction is only for major mergers based on the methodology used in this comparison (Conselice 2003). In this realization it is the WDM models which do better in matching the observations than the CDM models where a better agreement with the observed merger fraction is seen, in particular at higher redshifts. A caveat of this is that these merger data can be improved with more modern measurements. It should be stressed that Figure 7.1 uses imaging data and will be vastly improved in the future with new imaging missions such as Euclid and to some extent Rubin and Roman telescopes. It will also help in the future to have more simulations with different forms of dark matter, including its temperature and mass, to test how the merger history can be distinguished within different underlying dark matter particle properties.

7.3 Models of Galaxy Formation

As we have shown already in this book galaxy morphology and structure are new parameters in which to compare with cosmologically based galaxy formation models. These types of observations can also be used to test the physics inserted into cosmologically based models, including features such as star formation, AGN feedback, and supernova (Silk & Mamon 2012). These types of models through realizations such as Illustris and Eagle are becoming extremely sophisticated and realistic in terms of galaxy structure (Dickinson et al. 2018).

7.3.1 Testing Galaxy Formation Processes

Galaxy formation models involving the distribution of stars were first developed to explain the structures of galaxies, namely the properties of bulges and disks, and the relative ages of the stars in these components (Eggen et al. 1962). It was Eggen et al. (1962) who claimed that the physics of galaxy formation could be derived based on the structure of the Milky Way and the velocities of stars in its various components.

From that time forward, the default assumption within galaxy formation models was that galaxies formed like stars in a relatively rapid collapse. It makes sense to take this as a first assumption, as we know in general that stars form by a collapse, rather than hierarchically. However, as discussed in Chapter 5 we know that this assumption is incorrect and that galaxies form a large fraction of their mass through hierarchical processes such as mergers.

Galaxy formation models have also grown in tandem with cosmological simulations. In fact, it is cosmological simulations that predict how structure in the universe formed overall. Starting in the 1980s the first really detailed computer simulations of structure formation showed that a universe dominated by CDM matched observations of galaxy clustering on large scales (Davis et al. 1985) much better than other cosmologies, such as hot dark matter. This was important as the neutrino is a natural candidate to be the dark matter, but because it is a fast moving particle it can be ruled out, as we know that the universe must form through some type of CDM or WDM particle. Within this framework of CDM galaxy assembly should also be hierarchical (Blumenthal et al. 1984) in addition to the form of large scale structures. This idea that galaxies form in mergers is a fundamental prediction which is just now starting to be tested whereby direct comparisons between merger observations and theoretical predictions can be made (e.g., Bertone & Conselice 2009; Jogee et al. 2009; Hopkins et al. 2010; Lotz et al. 2011; Duncan et al. 2019).

There are many types of simulations in which predicted galaxy structure and morphology can be directly compared with observations of galaxies. The most direct is to compare the properties and structural features of distant galaxies to hydrodynamical models. There now exists many simulations that can be used to predict the properties of the galaxy population, and how it evolves. Models such as these are largely successful when predicting basic properties of nearby galaxies, such as their luminosities, masses, colors, and star formation rates. These models also compare well when they try to predict the form of scaling relationships of galaxies, such as the Tully–Fisher or size-mass relations. Historically, however, there have been problems with these models in predicting the abundances of low and high mass galaxies (e.g., Conselice et al. 2007; Guo et al. 2011). In addition to this, there are also significant mismatches when examining the evolution of merger histories compared with theory.

Ultimately, what we want from simulations is to go beyond reproducing what is already known about the galaxy population and to have the models come up with predictions and measurements that can be tested with data. Therefore, producing images of galaxy structures is one of the next big steps in this process. Early simulations of these images showed that the components of galaxies—namely bulges and disks—could be explained by a combination of mergers and accretion events (e.g., Steinmetz & Navarro 2002). These early studies argued that the Hubble type of a galaxy is not an aspect that survives for long periods of cosmic time, but can evolve. In a similar study, Governato et al. (2007) demonstrated that simulated disk galaxies can have properties that match the morphological, structural, and kinematic features of nearby disks. One caveat to some of these simulations is that they are not run in a cosmological context. The reason is that it is very difficult to predict the formation of galaxy morphology in simulations, and we are just now able to do this to some extent. In a real sense, this will be the ultimate test of galaxy formation models in the future.

As discussed in Chapter 4 one of the most commonly seen properties in high redshift star forming galaxies are large clumps of star formation within disks. A major question is how these clumps form, evolve, and their role in the formation of other galaxy components such as bulges and AGN and mass assembly. One example of a theory to explain this is Bournaud et al. (2014) who examined this problem and computationally determined how clumps with stellar masses of $M_* = 10^{8-9}$ $M_\odot$

evolve inside gaseous disks. One of the major question concerning these features is whether these clumps dissipate over a few Myrs or so, which is the dynamical time-scale of these clumps, or if they regenerate and survive over long periods of time. Bournaud et al. (2014) show that these clumps can last for about 300 Myr by acquiring new gas from its disk. However, some mass in these clumps can be lost through tidal effects. The question is how do these clumps relate to the formation of bulges and their overall formation, if at all. These simulations show that there is enough time for these clumps to migrate to the center of their host galaxy which can then fuel the AGN or merge with other clumps to form a proto-bulge. These simulations show that these clumps may provide a significant route for galaxies and their components to form. Theory thus shows that both inside-out and outside-in formation are predicted to occur within the galaxy population. What remains to be seen is whether one of these mechanisms is dominant, and the relative role of both in forming galaxies. This will most certainly have to be determined observationally in the future.

7.3.2 Merger History Predictions

One of the ways to further test models and simulations is to investigate how well CDM models can reproduce the merging process using hydrodynamical and the so-called semi-analytical methods (e.g., Bower et al. 2006; Guo et al. 2011). As mentioned, this is one of the ultimate ways in which to test the theory of galaxy formation. A comparison with measured major merger fractions is shown in Figure 7.3 at two different stellar mass ranges. Plotted as the thin solid dark blue line toward the lower part of each diagram is the prediction for the major merger fraction for galaxies from the Millennium simulation (Bertone & Conselice 2009). More recent comparisons are shown in Figures 5.4 and 5.5 in Chapter 5. In those figures the Eagle simulation is used to predict the merger fraction for galaxies. What can be seen is some agreement, but that overall the merger fraction predictions are significantly and consistently higher than the observations. This demonstrates the pattern seen in modern simulations predicting too many galaxies in mergers, or at least major mergers, over the epochs of cosmic time where we can make these measurements and comparisons.

This is a consistent theme in the published literature regarding how mergers in simulations compare with the observational data. There are simulations which are used to compare different aspects of the observations of mergers with predictions from theory. Maller et al. (2006) present cosmological hydrodynamical simulations of mergers for similar mass galaxies of a few times 10^{10} $M_{\odot}$, finding that the merger fraction predictions are not correct or consistent with each other, and that more simulation work should be focused on this critical aspect of the galaxy population.

It is possible to predict the merger history of observed galaxies using other methods such as the abundance matching technique (e.g., Stewart et al. 2008). This is a relatively new method whereby observed galaxies are matched to halos in models through their comparative abundances, or number densities. In simple terms,

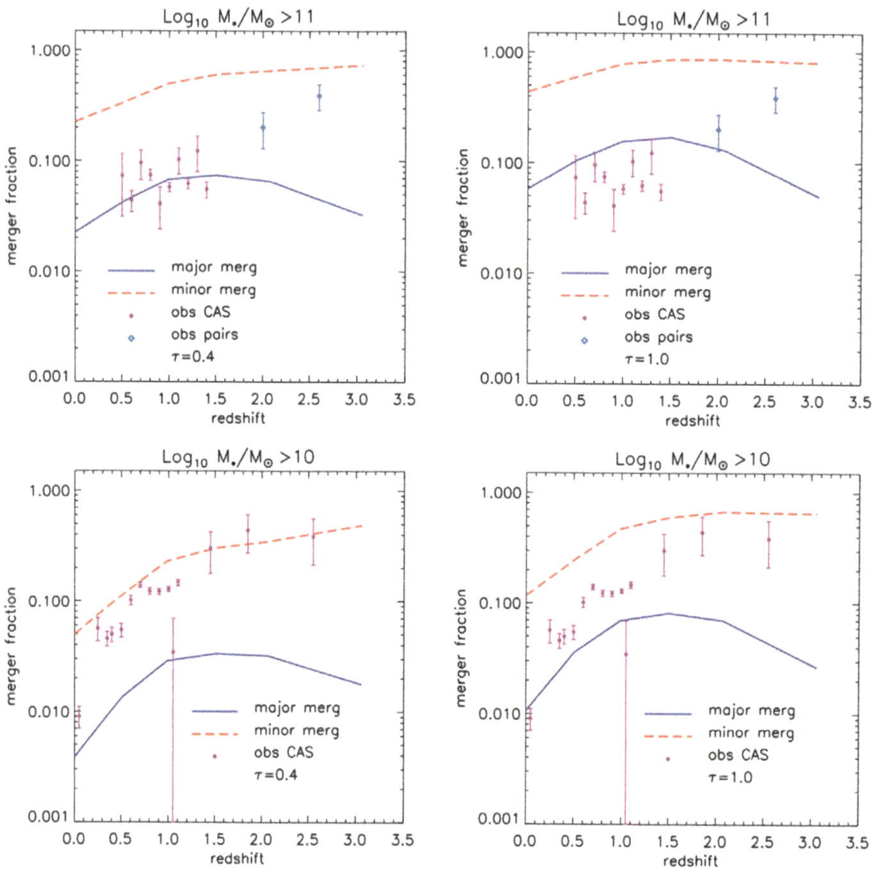

Figure 7.3. The merger predictions from the millennium simulation compared with CAS asymmetry derived merger fractions (Bertone & Conselice 2009). This figure shows that merger fractions are predicted to be often different than what is typically observed. This continues when comparing mergers to pairs through more model simulations such as Illustris, although the differences have become less acute with more up to date simulations (see text).

the rank order of galaxies based on stellar mass is matched 1-to-1 with halos produced from a simulation. Once this is done, then evolution can then be inferred by later evolution of both. As an application of this method Hopkins et al. (2010) predict the merger rate and fraction for galaxies between $M_* = 10^{10-11}\ M_{\odot}$. While the merger fractions from Hopkins et al. (2010) are higher than those from CDM models, they are still lower than the observations (see also Jogee et al. 2009 and Lotz et al. 2011 for further discussions).

Minor mergers are more common through the universe's history but they are really only just being measured recently (e.g., Bluck et al. 2012; Duncan et al. 2019). What is found when comparing minor merger fractions to simulations is shown in Figure 7.4. Here it can be seen that the predictions agree fairly well, but that the observed fraction of mergers is significantly, and consistently higher than the theoretical predictions from the Illustris simulation (Duncan et al. 2019). Overall,

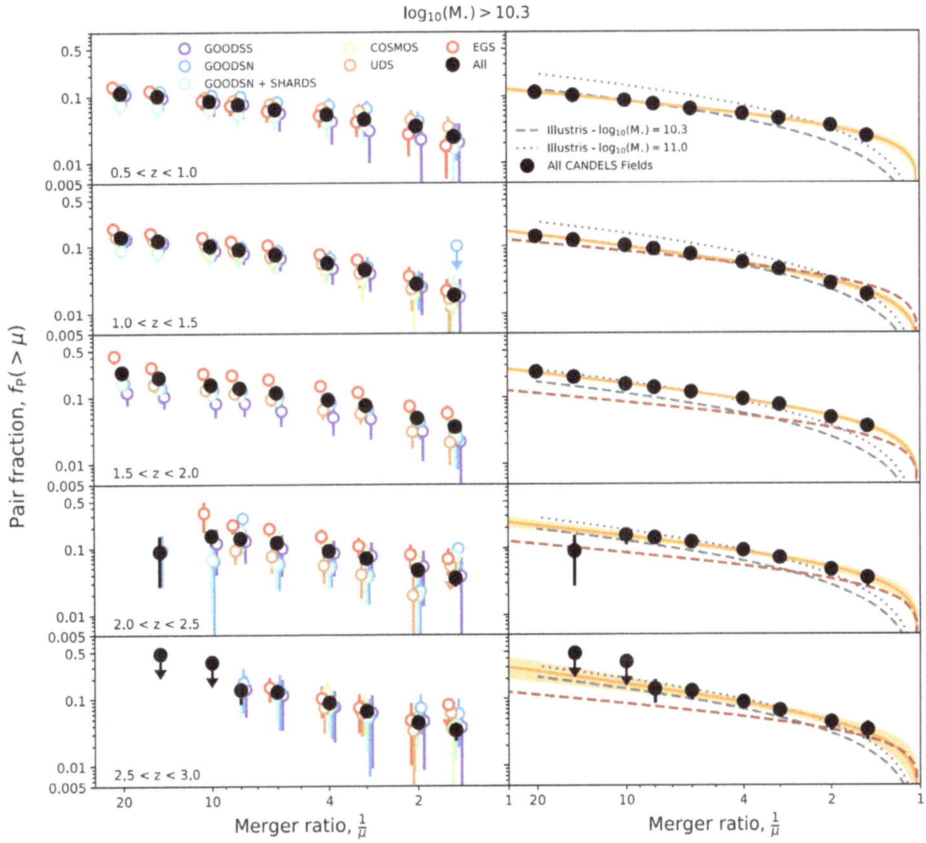

Figure 7.4. Minor merger evolution compared with predictions from the Illustris simulation. Overall, there is a better agreement compared with the major mergers at these redshifts, but in general the observations find a higher merger fraction for minor mergers than what is predicted in this simulation. Reproduced from Duncan et al. (2019).

this prediction and comparison to data for galaxy mergers and structure is a major way to determine the processes driving galaxy formation. This is an area where future work is certainly needed and will be very fruitful.

7.3.3 Future of Galaxy Formation through Mergers

The future is very promising for using galaxy structures and mergers for tracing the galaxy formation process, and the use of it for deriving properties of galaxy evolution and cosmology. There is still much about galaxy mergers, including their basic measurements, that we do not know or have measured to high accuracy. Also, using mergers as a cosmological tool has yet to be exploited in any significant way. Different methods for finding mergers are also in their infancy, as is studying mergers in the context of gravitational wave experiments.

Observationally, we still do not have a firm understanding of the merger history as a function of time to an accuracy similar to the star formation rate. This is mostly due to the relatively low level of research activity to date studying mergers as compared to star formation. Major progress can therefore still be made. There are still many uncertainties due to the lack of known spectroscopic pairs of galaxies at different redshifts. This is due to the fact that most galaxies surveys are done over large areas such that galaxies close together are very hard to obtain spectroscopy for in the same exposure, as these surveys are not optimized to find small-scale galaxy structures. Therefore, we have to use techniques such as photometric redshifts or morphologies, both of which have their downsides as explained earlier in this book. Obtaining spectroscopy for a large number of apparent pairs will go a long way toward having a precise measurement of the galaxy merger history through cosmic time, as well as its effect on star formation and AGN triggering. Thus, to make progress in this direction will require both observations and theory.

References

Bertone, S., & Conselice, C. J. 2009, MNRAS, 396, 2345
Bluck, A. F. L., Conselice, C. J., Buitrago, F., et al. 2012, ApJ, 747, 34
Blumenthal, G. R., Faber, S. M., Primack, J. R., & Rees, M. J. 1984, Nature, 311, 517
Bournaud, F., Perret, V., Renaud, F., et al. 2014, ApJ, 780, 57
Bower, R. G., Benson, A. J., Malbon, R., et al. 2006, MNRAS, 370, 645
Carlberg, R. G. 1991, ApJ, 375, 429
Conselice, C. J. 2003, ApJS, 147, 1
Conselice, C. J., Bundy, K., Trujillo, I., et al. 2007, MNRAS, 381, 962
Conselice, C. J., Bluck, A. F. L., Mortlock, A., Palamara, D., & Benson, A. J. 2014, MNRAS, 444, 1125
Davis, M., Efstathiou, G., Frenk, C. S., & White, S. D. M. 1985, ApJ, 292, 371
Dickinson, H., Fortson, L., Lintott, C., et al. 2018, ApJ, 853, 194
Duncan, K., Conselice, C. J., Mundy, C., et al. 2019, ApJ, 876, 110
Eggen, O. J., Lyden-Bell, D., & Sandage, A. R. 1962, ApJ, 136, 748
Governato, F., Willman, B., Mayer, L., et al. 2007, MNRAS, 374, 1479
Guo, Q., White, S. D. M., Boylan-Kolchin, M., et al. 2011, MNRAS, 413, 101
Hopkins, P. F., Croton, D., Bundy, K., et al. 2010, ApJ, 724, 915
Jogee, S., Miller, S. H., Penner, K., et al. 2009, ApJ, 697, 1971
Lotz, J. M., Jonsson, P., Cox, T. J., et al. 2011, ApJ, 742, 103
Maller, A. H., Katz, N., Kereš, D., Davé, R., & Weinberg, D. H. 2006, ApJ, 647, 763
Silk, J., & Mamon, G. A. 2012, RAA, 12, 917
Steinmetz, M., & Navarro, J. F. 2002, NewA, 7, 155
Stewart, K. R., Bullock, J. S., Wechsler, R. H., Maller, A. H., & Zentner, A. R. 2008, ApJ, 683, 597
Turner, M. S., & Riess, A. G. 2002, ApJ, 569, 18

Chapter 8

The Present and Future

The future of any field is often the most important thing about it in terms of interest and often relevance. If a field is researched out and there is not much left to discover, it becomes part of knowledge. However, for researchers it is always desirable for open questions to still exist, and that is certainly the case for understanding galaxy structures and how they evolve, and what these structures mean and imply. In fact, we are really just starting to understand the structures of galaxies, and there is much to learn both in terms of more detailed observations as well as for probing different and new yet to be discovered galaxies.

Although much has been learned about galaxy structure and morphology over the past century there are still many things which are not known, and aspects of galaxies that have yet to be investigated in great detail with available facilities. There will be great advances in this once the next generation of telescopes are online, starting with the James Webb Space Telescope (JWST) in 2021, but also including the Vera Rubin telescope, the Euclid Space Telescope, the Nancy Roman Observatory, and the Square Kilometer Array.

Furthermore, most of this book has focused on the optical and near-infrared properties of galaxies, tracing various aspects of their stellar distributions. However, the structure of galaxies in other wavelengths can probe other aspects of galaxies, such as their gas content and it is becoming possible to even study their dark matter structures. Another area which looks very promising is ultra-high resolution imaging that can probe the smallest scales in galaxies which we cannot reach today. This will be especially important for understanding the central regions of galaxies where supermassive black holes exist, but also for deciphering the structures of unresolved regions, such as star-forming clouds among other regions. Overall, this field is very promising for its ability to make major new discoveries. Below we give just a flavor for the progress being made toward looking into these issues and where future directions can take us.

doi:10.1088/2514-3433/abb602ch8

8.1 Extended Diffuse Light in Galaxies

There have been major advances in our understanding of resolved galaxies over the past 20 years, both through nearby galaxy imaging with, e.g., the Sloan Digital Sky Survey (SDSS) and using the Hubble Space Telescope (HST) at higher redshifts. At the same time, we have learned a great deal about galaxies and their properties through multi-wavelength studies to characterize everything from their ongoing star formation rates to the presence and activity of central massive black holes.

Imaging of galaxies is one of the main ways this is done and is the primary focus of this book. Historically this has involved looking at and examining the brightest portions of galaxies—namely its main areas of stellar mass and star formation. To date, existing programs to examine distant galaxies have not included many intrinsically faint portions of massive galaxies or low mass galaxies as these studies were designed to examine either shallow large areas with imaging of the nearby universe, or in very deep but small area imaging designed to study distant galaxies. However, there are other features and locations within galaxies that are important, and display information about their history and formation. This includes the outer portions of light in galaxies as well as imaging of the circumgalactic medium (CGM) discussed in the next section. The CGM is the outer halo portion of a galaxy which has been studied mostly in terms of its gas and in some cases the amount of light which is emitted from this gas. Both of these topics are relatively new and are just now being studied in detail.

The diffuse outer light in galaxies is also something that can be seen and observed through deep imaging. Typical exposures of nearby galaxies are still quite often relatively short, but by going deeper it is possible to see structures that otherwise would not be visible on typical depth exposures. One example of this is shown in Figure 8.1 which is a deep Canada–France–Hawaii Telescope (CFHT) image of the famous galaxy M51 and its companion/interacting galaxy. As seen in this image there is an extensive amount of diffuse light surrounding these two interacting galaxies which, if studied in detail, likely can reveal the origin of these structures and the galaxies themselves. In fact, these are very useful structures as they often have survival times which are very long, much longer than in the inner portions of galaxies.

One way to see this is through calculating the timescale for structures to erase themselves through mixing or dissipation. This timescale is in general given by the crossing time which is calculated as the size of the structure (R) divided by the velocities of the stars at those radii (σ), or $\tau_{\rm cross} = R/\sigma$. Thus for a given velocity σ the crossing time, and therefore the time for structure to become dissipated, is longer in the outer portions of galaxies. Detailed simulations confirm that these tidal debris features will survive for several Gyr (e.g., Naab et al. 2006). Thus, in many ways by looking at this outer light we are getting some idea about the formation mechanisms, and interestingly the past history, for galaxies.

One of these features is to account for accreted galaxies that have merged with the central galaxy. These events will leave residual light remaining that can reveal properties of past construction events, such as the merger of satellites galaxies to

Figure 8.1. M51 as observed with deep CFHT imaging. As can be seen, this system is composed of two different galaxies that are interacting. As such, there is a large amount of diffuse light in the outskirts of these systems which is likely partially due to the interaction dislodging stars from these galaxies. A typical image of this system would not show such features; very deep and high-quality imaging is needed to see them. Credit: image by Jean-Charles Cuillandre (CFHT) & Giovanni Anselmi (Coelum), Canada–France–Hawaii Telescope (CFHT)/Coelum. © 2020 CFHT. https://www.cfht.hawaii.edu/HawaiianStarlight/trailer.html

form the larger structures of the galaxies. Also in cases such as M51 (Figure 8.1), these outer light structures can be used to see the interactions between galaxies through the amount and form of light which is being disturbed by these tidal events. If one were to take a shallow image of M51 it would not be possible to see these outer tidal events and it would appear that there were just two nearby galaxies that were relatively undisturbed.

Some telescopic programs have studied low luminosity galaxies and halo light in small targeted investigations (e.g., Martinez-Delgado et al. 2010; Duc et al. 2015; Ruiz et al. 2015). What we have learned is that there is a significant low surface brightness component to galaxies that is still in its infancy in terms of being characterized. Duc et al. (2015) found that 1/2 to 2/3 of nearby massive galaxies contained outer halo light in the form of tidal tails (e.g., Figure 8.1). Thus many massive galaxies will contain tidal features in their outer portions, which are likely due to mergers and accretion of satellite galaxies over the past few Gyr (e.g., van Dokkum 2005; Duc et al. 2015).

What is the prospect for the future of this field? The SDSS survey plus GAMA survey has obtained the deepest and most thorough survey of nearby faint galaxies to date, yet has only studied a few thousand of these systems over several hundred deg^2 (e.g., Kelvin et al. 2014). Surveys such as KiDS has a deeper depth than SDSS, yet the area covered is very small at 250 deg^2. Likewise, the halos of galaxies, beyond a few scale lengths are also understudied due to their faint low surface brightness nature which requires relatively deep exposure times to study and measure. Less than a few hundred galaxies have had their outer light studied (e.g., van Dokkum 2005; Duc et al. 2015). However, this will change dramatically in the next few years with the advent of new telescopes that will acquire deep imaging over large areas of the sky. This includes the Euclid space telescope, as well as the Vera C Rubin Telescope, which will obtain imaging such as this but for up to tens of thousands of nearby galaxies. This will really allow us to put constraints on these features and derive what they imply for the formation and evolution of galaxies.

8.2 Emission from the Circumgalactic Medium

Another relatively new feature of galaxies is the ability to see the structures in the form of gas surrounding them, such as the circumgalactic medium (CGM). The CGM is located in the outer portions of galaxies and is now one of the main areas of recent interest for absorption line studies. Absorption line studies are a relatively mature field of astronomy whereby gas clouds in the intergalactic medium can be studied through absorption lines as seen in the light from quasars behind these gas clouds (e.g., Sargent et al. 1980). When these systems were first studied what they located where large apparent reservoirs of gas that perhaps galaxy formation would later occur from. As techniques improved it became possible to find less dense gas systems such as Lyman-limit systems with lower column densities than the large column density damped Lyα absorption systems.

Recently we have been able to study the gas surrounding known galaxies in the form of this CGM. This is important for galaxy structures as this gas in the CGM occupies a significant fraction of the gas mass within galaxies. Probing the structure of gas in the outer parts of galaxies with just absorption line systems is very difficult as a single sight-lines probe only a small part of a system, and naturally to examine the entire structure we need to have multiple sightlines all going through different aspects of the galaxy's structure, which is impossible to do currently. However, it is possible that we can see this light originating from the CGM due to processes within the gas that produces light, such as fluorescence whereby the atoms are excited and drop to a lower energy, producing photons that we can detect.

For example, the Keck Cosmic Web Imager has been able to image this surrounding light around high redshift QSOs in Lyα emission (Cai et al. 2019). Many of these systems have extended emission up to 50 kpc, which is much larger than the sizes of the optical light originating from these galaxies. This light is also well fit by a power-law distribution. Some of these systems also have irregular distributions of this light, which is more commonly seen at $z = 2$ than at higher redshifts. This type of Lyα emission is also seen with MUSE on the VLT for even higher redshift QSOs at $z \sim 6$ (e.g., Farina et al. 2019). This diffuse outer emission is also often found to be irregular in shape.

8.3 Machine Learning Approaches

In the last few years a new epoch in the study of galaxy structures has been opened up by the possibilities of machine learning and deep learning applied to galaxies. For many years these methods were applied using tools such as artificial neural networks and hierarchical tree methods, but the outcome was never ideal and left much to be desired in terms of accuracy for classification (e.g., Huertas-Company et al. 2009).

Machine learning for the classification of objects is quite old now, starting from Fukushima et al. (1975, 1980). However, it took some time before machine learning was applied to astronomy with some early papers including Odewahn et al. (1992) and Weir et al. (1995), which were predominately interested in using early machine learning techniques to separate galaxies from stars in imaging surveys. A very natural expansion of this is to use these methods to classify galaxies into different morphological types, with the division of the Hubble sequence being one of the main goals. This was attempted by Storrie-Lombardi et al. (1992) using Artificial Neural Network algorithms, but the accuracy was only 64% of correctly classified systems. This was later improved by Lahav et al. (1996) to obtain better results. These early machine learning techniques used a Bayesian methodology to determine the posteriori probability for each possible morphological type for each galaxy. These methods using neural networks worked fairly well with up to a 90% accuracy, but were never as good as visual classification (e.g., Banerji et al. 2010). This was later expanded using support vector machines (e.g., Huertas-Company et al. 2009), as well as other similar methods such as Classification Trees and Random Forest methods (Sreejith et al. 2018).

The use of computers for determining galaxy morphology received a major boost and became practical and superior even to visually based methods with the advent of Convolutional Neural Networks (CNN), a form of "deep learning" (e.g., Dieleman et al. 2015; Huertas-Company et al. 2015; Huertas-Company et al. 2018; Domínguez Sánchez et al. 2018; Walmsley et al. 2020; Cheng et al. 2020a; Figure 8.2). This came about with the direct use of computer science methodologies applied to astronomy as influenced by a competition to determine a machine-based

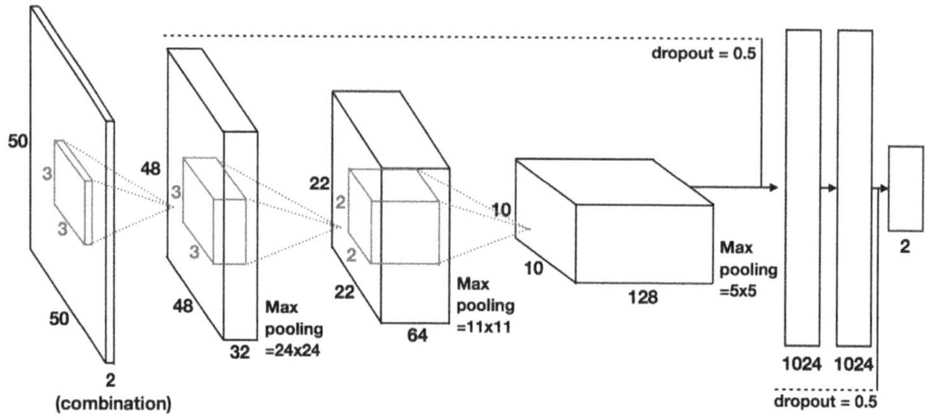

Figure 8.2. The schematic overview of the architecture of CNN from Cheng et al. (2020a). The architecture starts from an input image with size 50 by 50 pixels, then three convolutional layers (filter: 32, 64, and 128). Each convolutional layer is followed by a pooling layer. Two hidden layers with 1024 hidden units each are following the third convolutional layer. One dropout ($p = 0.5$) follows after the third convolutional layer and the other follows after the second hidden layer. For this particular system there are two outputs after this CNN classification: "ellipticals" and "spirals."

way of measuring the morphologies of galaxies classified as part of the Galaxy Zoo project. It was Dieleman et al. (2015) who applied CNNs to the SDSS imaging, obtaining a nearly 99% accuracy in classification. The entire idea behind CNN is to use the entire image itself, and not "feature extract" which was necessary with the neural network approach. This way the entire system's image is used, and no predetermined (by the user) features are fed into the system such as galaxy measurables, some of which are described in earlier chapters. Immediately, it was clear that the way forward with galaxy classification into types is through this CNN, and other methods, of using machine learning (e.g., Cheng et al. 2020a). In fact, these methods are now so powerful that machine learning through CNN is able to find mistakes and misclassifications carried out by Galaxy Zoo due to the low resolution of the SDSS imaging data (e.g., Cheng et al. 2020a) and have an accuracy of greater than 99% (Figure 8.3). Future approaches will likely use a combination of human classifications and those of a machine to retrain difficult and more unique cases with extra human classifications (e.g., Beck et al. 2018). This is likely how actual machine learning methods for future large surveys such as Euclid and Rubin will have to be carried out.

Naturally, we want to be able to examine more than just the Hubble morphology when we examine the structures of galaxies. There are other issues that CNN and machine learning can be applied to including strong lensing identification, which works fairly well with CNN training and even unsupervised learning (e.g., Petrillo et al. 2017; Lanusse et al. 2018; Cheng et al. 2020b). There are also more than just Hubble types to galaxy structure, and attempts to find merging galaxies with deep learning have also been carried out (e.g., Bottrell et al. 2019; Ferreira et al. 2020), but in general are not yet at the success level as binary classifications between

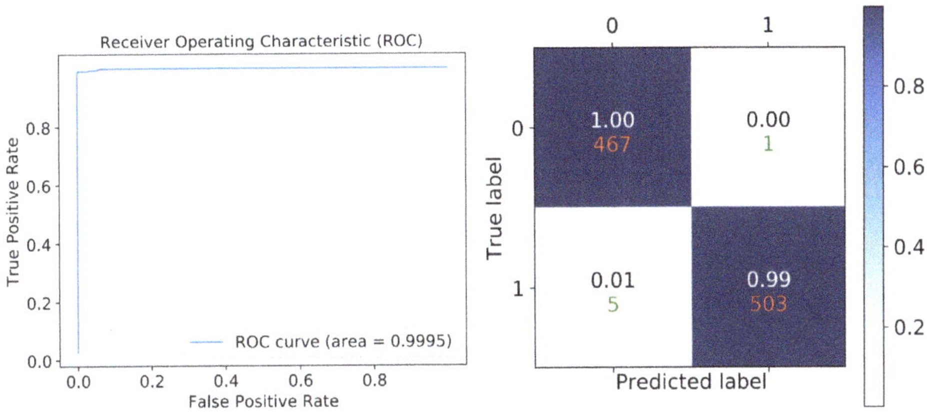

Figure 8.3. One of the best testing results for machine learning classifying ellipticals and spirals from (Cheng et al. 2020a). The right panel shows the Confusion matrix whereby the "0" are for ellipticals and the "1" represents spirals. The color bar shows the fraction of each true label (Galaxy Zoo), and the number shows the corresponding fraction. The left panel shows the corresponding receiver operator curve (ROC) of this testing result. When these ROC curves approach the 0,0 point, an ideal training and classification of the data has been achieved.

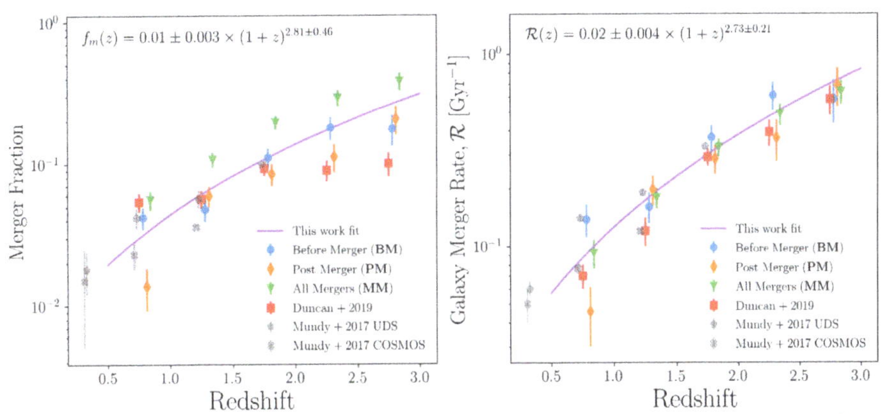

Figure 8.4. Plot showing the merger history of galaxies up to $z = 3$ as measured through deep learning training of the Illustris simulation applied to CANDELS imaging. Plotted on this figure are previous merger fractions measured with pairs of galaxies as shown in the caption. There are various types of mergers measured in this study, including "before mergers," "post-mergers," and a combined "all mergers" in green. The line is the best fit through the machine learning points, and the shading shows the range of possible fits. Reproduced from Ferreira et al. (2020).

ellipticals and spirals. However, despite this, we are now able to measure the merger history using machine learning to the same level as we can use pairs of galaxies and structural measurements directly (see Figure 8.4). Studies are also ongoing to determine how phases of merging can be distinguished within deep learning training. One way to do this is to apply machine learning trained on simulations of the merger process whereby each phase of a merger is categorized and the computer is trained to learn how to find each particular stage. This training is then applied to data to determine the actual merger history for a collection of galaxies. Methodologies like this will greatly improve our knowledge of the merging process whereby we will soon be able to study the entire history in much more detail. This may reveal how mergers are correlating with other processes in more detail, such as star formation triggering, star formation truncation, as well as AGN fueling to an even better accuracy than we currently can using our own predetermined merger phases (see Chapter 6).

New techniques in machine learning and deep learning will clearly advance our understanding of galaxy structure and how to classify objects into categories. However, there are natural liminations to this process as clearly the ability to make something useful with this is limited by our ability to known which classifications are the best to use and to be sure that the training set is correctly assigned. While machine and deep learning are revolutionizing the study of galaxy structure, we are only in many ways still reproducing the original classification schemes that have trouble representing real physical meaning (see Chapter 2). What we would like to be able to do is let the machine classify galaxies into different types in which it has found by itself as the most natural categories, or "clusters"—this is the unsupervised method which is similar to the PCA analysis discussed in Chapter 2, and for which is already providing unique insights into how to classify galaxies into "learned" classes (Hocking et al. 2018; Uzeirbegovic et al. 2020). Another way to use machine learning is to use "regression" to determine values of galaxy parameters, such as the CAS parameters. This is such that the machine is able to retrieve these values by using the image as input. The benefit of using a machine learning method for this rather than directly measuring it is that the machine is up to a 100 times faster, and will be the only suitable way to measure these features in the upcoming era of big galaxy surveys.

8.4 The James Webb Space Telescope (JWST)

Since the beginning of large-aperture ground and space telescopes, astronomers have been pushing further back in time to when we can study galaxies. The Hubble Space Telescope (HST), along with ground-based telescopes such as the Very Large Telescope (VLT) and the Keck Telescopes, have provided a great advance over the past 20 years, to the point where we can now probe galaxies to within a billion years after the big-bang. Despite this, no current observational facilities are able to probe the earliest galaxy and star formation in the universe. The challenges are in finding the first galaxies, identifying them from contamination, and making accurate measurements of luminosity and mass functions, as well as measuring resolved structural information which as we have discussed, can tell us about the formation mechanisms of these galaxies. These measurements will tell us the physics of early galaxy formation, as the quantitative evolution of these quantities is predicted differently in models that make different assumptions. By comparing observations with these models we can derive the physics behind the earliest formation of galaxies, something that we cannot do at the time of writing.

As we have described in the earlier chapters of this book, at redshifts lower than $z = 3$ we know a great deal about the formation and evolution of galaxies. We know that the epoch at $z \sim 2 - 3$ is where the peak of star formation occurred (e.g., Madau & Dickinson 2014), and where there is an already diverse galaxy population with a range of masses, structures/sizes, morphologies, and star formation histories (e.g., Conselice 2014). The bulk of the stellar mass in the universe's galaxies formed after $z = 3$, however the unknown initial seeds of this formation at $z > 3$ determined its later evolution and the final forms we see today. However, we have really only just begun to study these systems. To make progress toward understanding the earliest epochs, which sets the stage for the galaxies we find today, we need to probe $z > 3$ to decipher how this initial formation itself took place, including the structures of these early galaxies.

Very likely major clues toward our ultimate knowledge of galaxy formation and evolution at the earliest epochs of the universe will be revealed in the next five years with the James Webb Space Telescope (JWST), launching in 2021. The broad scientific goals of JWST are to determine how galaxies, stars, and planets have formed and evolved. JWST will be one of the most unique scientific instruments ever constructed. It will have an aperture comparable to the largest optical telescopes on the ground, but will also have the advantage of being in space. Since JWST will be beyond the Earth's atmosphere, it will not have to deal with distortions that cause all ground-based images to appear blurry due to refraction of light due to atmospheric disturbances.

JWST will be an infrared telescope operating in the wavelength range 0.5–28 μm. Its large aperture, red wavelengths, and extreme depth capabilities, even compared with HST, allows JWST to answer the most fundamental questions concerning the process of the earliest galaxy formation that will not be achievable by any other telescope being designed, or built, in at least the next 20 years and likely much longer.

One great advantage of JWST for early galaxy studies is that it will allow us to probe the rest-frame optical light of galaxies at high resolution during the first great epoch of galaxy mass assembly between redshifts $3 < z < 7$. This is not possible to do with HST, as it cannot probe photons redward of the observed H-band (1.6 μm). Thus, JWST will allow us to make great progress in understanding the galaxy formation history of the earliest galaxies from when the initial seeds of galaxy formation are planted. This can be seen in the factor of 10 increase in the star formation rate from $z = 7$ to $z = 3$, which rapidly builds up the masses of these galaxies during this epoch. A simulated image of what JWST will see in a deep field is shown in Figure 8.5.

Exactly how this formation occurred and its role in producing the structures of galaxies is still unknown, but will be one of the major findings of JWST. This will be addressed with JWST NIRCam imaging out to 5 μm, a wavelength where no high resolution data has been obtained yet. We know that some spheroids, disks, and bulges are formed by $z = 3$, thus with the range of NIRCam filters and with this high resolution NIRCam imaging we will investigate the onset of the formation of bulges and disks using methodologies that we developed for $z < 3$ galaxies using HST data (e.g., Margalef-Bentabol et al. 2016).

The JWST which while will not typically have a much higher resolution than HST, will none the less be focused on the reddest wavelengths where we have very

Figure 8.5. A simulation of what a JWST deep field will look like. Compare this with the deepest HST data discussed in Chapter 4. Credit: STScI. CC BY 4.0. https://spacetelescope.org/images/jwst_simulation_halfsize/.

little high resolution information with existing and past facilities. This is due to the fact that longward of $\lambda > 3$ μm it is not possible to obtain resolutions better than a few tens of arcsec at the moment with, e.g., Spitzer data. However, if we want to examine galaxy structures in the rest-frame optical at redshifts higher than about $z > 3$, then we need to have high resolution imaging at these mid-IR wavelengths.

8.5 Non-Optical Structure and Morphology

Another aspect of the future of the field of galaxy structures and morphologies is to examine these features at wavelengths other than optical/UV/NIR and perhaps even to go beyond photons and measure structures in terms of stellar or even total masses dominated by dark matter. Since the discovery of gravitational waves, perhaps we will even be able to one day make a map of a galaxy through gravitational waves, although this is something that will likely not happen until at least the end of the 21st century.

One of the most obvious things to first examine is how the structures of galaxies change as a function of wavelength. It is clear that within the optical wavelength range from about 4000 Å to 9000 Å galaxies look very similar both in terms of their overall structural appearance and when this structure is quantified (e.g., Windhorst et al. 2002; Taylor-Mager et al. 2007; Mager et al. 2018). However, this changes when viewing the structures of galaxies below what is commonly called the "balmer break" at 3646 Å, whereby galaxies appear very different. This is because at wavelengths shorter than this we are probing light in the ultraviolet which is dominated by young stars for the most part. For galaxies which are mostly composed of old stars, such as ellipticals, the structure and morphology of galaxies within the UV is very different than it is at longer wavelengths, and sometimes very little emission is present. Considering galaxies where there is a star-forming disk combined with an older bulge, the galaxy looks very different in the UV and optical, with very little UV flux in the center and a more prominent UV disk. In terms of non-parametric parameters we find that galaxies are less concentrated and more asymmetric in the UV (Mager et al. 2018).

There are of course other wavelengths in which galaxies emit light, including in X-rays as well as radio. Since HI, or the 21 cm line, emission traces neutral hydrogen, in many ways, it is competitive with optical emission for revealing a galaxy's structure. Both methods probe a significant fraction of the baryonic mass in galaxies, in particular for late-type systems dominated by star formation. There are a couple of things about HI observations that may make them even superior for tracing of structure in galaxies. The first is that galaxies are much larger in HI observations compared to when they are viewed in optical light (see Figure 8.6). The other is that galaxy interactions and mergers can be seen as disturbances of the HI distribution more readily than when viewed in optical light. This is also due to the longer timescales in the outer parts of galaxies as described earlier in this chapter for the diffuse optical light.

Holwerda et al. (2011a) was one of the first to investigate this issue quantitatively with CAS and other indices. They found that the asymmetry index and M_{20} were the two most promising for finding interactions using this neutral hydrogen imaging. Furthermore, it was shown that the life-span of these merger features in HI lasted much longer than those in the optical, with at least 1 Gyr before they dissipate (Holwerda et al. 2011b). With the future advent of telescopes such as SKA, understanding the structures of galaxies in HI gas will become even more important and will be a major aspect toward understanding galaxy evolution in terms of gas structures.

Recent advances in this area have been accomplished by the Atacama Large Millimeter/submillimeter Array (ALMA) observations of distant galaxies. With ALMA we can obtain high resolution imaging of distant star-forming galaxies up to $z \sim 6$ at typical resolutions of about 1 kpc. The emission seen from sources observed by ALMA is a measure of their dust emission and its distribution. This gives a

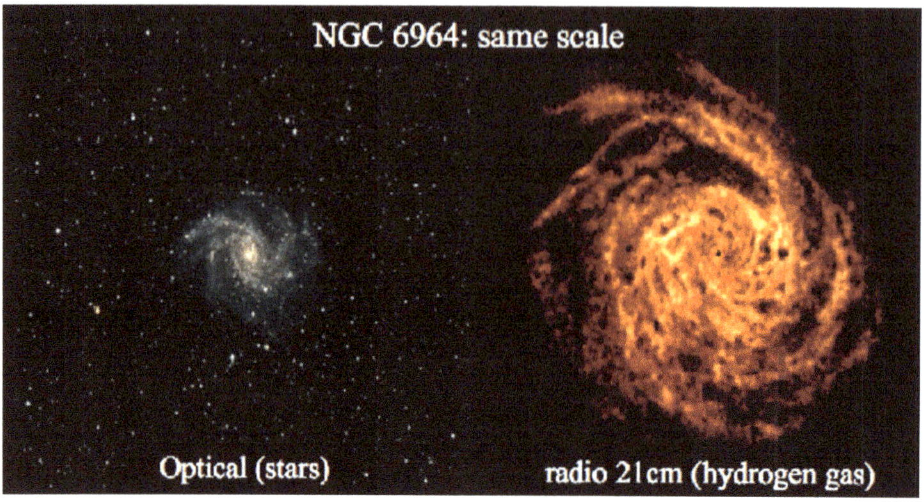

Figure 8.6. NGC 6964 as seen at the same scale, but imaged in the optical on the left and within 21 cm HI observations on the right. As can be seen the structure of this galaxy is very different between stars and gas, and when imaged in gas it is a much larger system. Credit: Rense Boomsma and Tom Oosterloo.

different view of galaxies than what is found in the optical (e.g., Gullberg et al. 2019). These images with ALMA can be analyzed in a similar way as we have done thus far for optical and NIR images. Most of these images have structures that can be fit with Sérsic indices of $n \sim 1$, and show indications of bulges in formation due to other structural information (Gullberg et al. 2019).

8.6 Very High Resolution Imaging

A very exciting area for the future is in high resolution imaging of galaxies and what that can tell us about the structures of these systems and their components. This is an area that is still rapidly developing and currently utilizes very-long-baseline interferometry (VLBI) stations across the world. One goal of this type of analysis is to image galaxies at ultra-high resolution in the near-infrared and optical, but this is very difficult given the wavelengths that must be probed to carry this out.

The Event Horizon Telescope (EHT; Event Horizon Telescope Collaboration et al. 2019a) is an interferometer the size of the Earth with stations in Hawaii, the Canary Islands, Chile, and the South Pole among other locations where radio and sub-mm telescopes are used to obtain high resolution imaging of, at the present time, the central parts of galaxies. The resolution of interferometers is given by λ/D, where D is the separation of the different telescopes used to produce the images. The value of D is the size of the Earth for the EHT, thus giving us some of the highest resolution of any images at 25 microarcseconds, which is by far the best resolution imaging obtained to date at any wavelength. This clearly will provide a new epoch in galaxy structure, whereby we can probe very small-scale features of galaxies in a way that we cannot currently do with optical light, for instance.

Thus far the EHT has released an image of the event horizon of the central massive black hole in the galaxy M87, and the shadow this produces. M87 was one of the first galaxies with a known supermassive black hole with a mass of about $6.5 \times 10^9\ M_{\odot}$, which is also one of the most massive known, with a mass similar to many dwarf galaxies. While it and other central massive black holes have been known about for decades, their presence has been inferred due to the activity of these black holes which we see as active galactic nuclei, or through kinematic measurements which give the mass of the black holes by examining how fast stars are moving very close to the centers of these galaxies.

The black hole in the center of M87 was chosen for this imaging experiment because it is so nearby and massive. The size of a black hole depends only upon its mass, and because this system is so close to us, it has the largest apparent diameter of any black hole, along with the black hole in the center of our own galaxy. When imaged, the size of this black hole could be measured, as well as the visible shadow the supermassive black hole made within the central part of the system (Figure 8.7). The EHT has been observing since 2006, adding data from different telescopes at 1.3 mm in the sub-mm, as they joined the project. Their conclusions based on the data is that the black hole imaged by EHT is consistent with a spinning Kerr black hole with the spin vector pointed away from earth (Event Horizon Telescope Collaboration et al. 2019b).

While imaging of the central massive black holes in galaxies is not quite the same as broad galaxy structure and how galaxies evolve, it does demonstrate that imaging at high resolution in at least some wavelengths will be possible for larger samples in

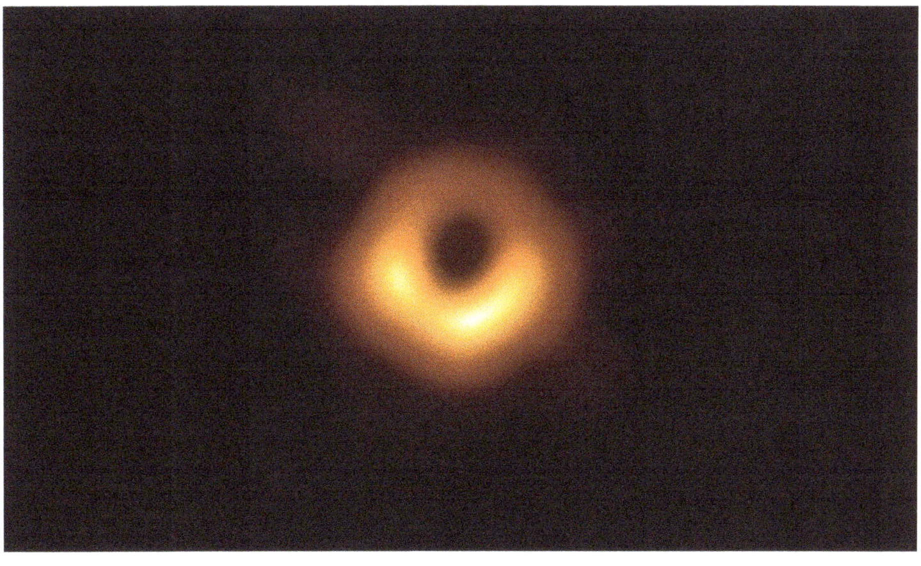

Figure 8.7. The Event Horizon Telescope (EHT) image of the center of M87, showing the shadow of the supermassive black hole which resides there. The total size is 42 microarcs and a new mass of this black hole was derived based on these observations. The brightness is in units of brightness temperature. Credit: EHT Collaboration. CC BY 4.0. https://www.eso.org/public/images/eso1907a/.

the near future. Black hole shapes and sizes are also part of galaxy structure, and in this sense these new EHT observations are probing a new part of parameter space in the study of galaxies. These techniques could be used in other studies such as the star formation or detailed structure of spiral arms, among other properties. This is still very early days for this type of high resolution imaging, but it will be possible to one day measure these properties in distant galaxies. This will allow for the changes in small scale structure in galaxies to be observed, opening up an entirely new parameter space of galaxy structural evolution studies.

8.7 CODA

As discussed throughout this book, the structures of galaxies is one of the key aspects for understanding galaxies and provides unique information on their formation and physical state. As we enter a new epoch in galaxy studies with new instruments such as Euclid, SKA, and JWST we will probe galaxies in a new way and reveal new structures. In many ways, the use of galaxy structure to understand galaxies, and the universe itself as a result, are still in their infancy. New approaches will have to be developed to deal with the big data that will arrive in the next decade and beyond with machine learning as well as the imaging of galaxies in unseen conditions in the early universe. We will also be able to image previously unseen portions of galaxies and will have to find new ways of quantifying this information. We have just started probing the very cores of galaxies, but the high resolution imaging of other parts of galaxies awaits future technologies and observing programs.

As we have demonstrated throughout this book, the field of galaxy structure is intimately linked with the development of galaxy formation. As discussed, some of the earliest observations on galaxies were to note their structure and morphological properties, which were some of the very first applications of systematic thinking applied to galaxies. This continues to be a major way in which we view galaxies and as we obtain new and different data we will examine galaxies in new ways, and structure will be at the forefront of that activity. Galaxy structural analysis has always been a major way in which galaxies have been studied, and this will continue to be the case. As we approach the future new data and techniques will appear as necessary, which makes the field of galaxy structural analysis an important one which in some ways is still just beginning.

References

Banerji, M., Lahav, O., Lintott, C. J., et al. 2010, MNRAS, 406, 342
Beck, M. R., Scarlata, C., Fortson, L. F., et al. 2018, MNRAS, 476, 5516
Bottrell, C., Hani, M. H., Teimoorinia, H., et al. 2019, MNRAS, 490, 5390
Cai, Z., Cantalupo, S., Prochaska, J. X., et al. 2019, ApJS, 245, 23
Cheng, T.-Y., Li, N., Conselice, C. J., et al. 2020a, MNRAS, 494, 3750
Cheng, T.-Y., Conselice, C. J., Aragón-Salamanca, A., et al. 2020b, MNRAS, 493, 4209
Conselice, C. J. 2014, ARA&A, 52, 291
Dieleman, S., Willett, K. W., & Dambre, J. 2015, MNRAS, 450, 1441
Domínguez Sánchez, H., Huertas-Company, M., Bernardi, M., et al. 2018, MNRAS, 476, 3661

Duc, P.-A., Cuillandre, J.-C., Karabal, E., et al. 2015, MNRAS, 446, 120
Event Horizon Telescope Collaboration, Akiyama, K., Alberdi, A., et al. 2019a, ApJL, 875, L1
Event Horizon Telescope Collaboration, Akiyama, K., Alberdi, A., et al. 2019b, ApJL, 875, L5
Farina, E. P., Arrigoni-Battaia, F., Costa, T., et al. 2019, ApJ, 887, 196
Ferreira, L., Conselice, C. J., Duncan, K., et al. 2020, ApJ, 895, 115
Fukushima, K. 1975, Biol. Cybern., 20, 121
Fukushima, K. 1980, Biol. Cybern., 36, 193
Gullberg, B., Smail, I., Swinbank, A. M., et al. 2019, MNRAS, 490, 4956
Hocking, A., Geach, J. E., Sun, Y., et al. 2018, MNRAS, 473, 1108
Holwerda, B. W., Pirzkal, N., de Blok, W. J. G., et al. 2011a, MNRAS, 416, 2401
Holwerda, B. W., Pirzkal, N., de Blok, W. J. G., et al. 2011b, MNRAS, 416, 2437
Huertas-Company, M., Tasca, L., Rouan, D., et al. 2009, A&A, 497, 743
Huertas-Company, M., Gravet, R., Cabrera-Vives, G., et al. 2015, ApJS, 221, 8
Huertas-Company, M., Primack, J. R., Dekel, A., et al. 2018, ApJ, 858, 114
Kelvin, L. S., Driver, S. P., Robotham, A. S. G., et al. 2014, MNRAS, 444, 1647
Lahav, O., Naim, A., Sodré, L., & Storrie-Lombardi, M. C. 1996, MNRAS, 283, 207
Lanusse, F., Ma, Q., Li, N., et al. 2018, MNRAS, 473, 3895
Madau, P., & Dickinson, M. 2014, ARA&A, 52, 415
Mager, V. A., Conselice, C. J., Seibert, M., et al. 2018, ApJ, 864, 123
Margalef-Bentabol, B., Conselice, C. J., Mortlock, A., et al. 2016, MNRAS, 461, 2728
Martinez-Delgado, D., Gabany, R. J., Crawford, K., et al. 2010, AJ, 140, 962
Naab, T., Jesseit, R., Burkert, A., et al. 2006, MNRAS, 372, 839
Odewahn, S. C., Stockwell, E. B., Pennington, R. L., et al. 1992, AJ, 103, 318
Petrillo, C. E., Tortora, C., Chatterjee, S., et al. 2017, MNRAS, 472, 1129
Ruiz, P., Trujillo, I., & Mármol-Queraltó, E. 2015, MNRAS, 454, 1605
Sargent, W. L. W., Young, P. J., Boksenberg, A., et al. 1980, ApJS, 42, 41
Sreejith, S., Pereverzyev, S., Kelvin, L. S., et al. 2018, MNRAS, 474, 5232
Storrie-Lombardi, M. C., Lahav, O., Sodre, L., et al. 1992, MNRAS, 259, 8P
Taylor-Mager, V. A., Conselice, C. J., Windhorst, R. A., et al. 2007, ApJ, 659, 162
Uzeirbegovic, E., Geach, J. E., & Kaviraj, S. 2020, arXiv: 2004.06734
Walmsley, M., Smith, L., Lintott, C., et al. 2020, MNRAS, 491, 1554
Weir, N., Fayyad, U. M., & Djorgovski, S. 1995, AJ, 109, 2401
Windhorst, R. A., Taylor, V. A., Jansen, R. A., et al. 2002, ApJS, 143, 113
van Dokkum, P. 2005, AJ, 130, 2647

www.ingramcontent.com/pod-product-compliance
Ingram Content Group UK Ltd.
Pitfield, Milton Keynes, MK11 3LW, UK
UKHW061857100926
8556IPUK00008B/910

9 780750 326681